Developmental Biology

— A Manual —

O.P. Jangir

Laboratory of Developmental Biology
Post Graduate Dept. of Zoology
Dungar College, Bikaner

AGROBIOS (INDIA)

Published by:
AGROBIOS (INDIA)
Agro House, Behind Nasrani Cinema
Chopasani Road, Jodhpur 342003
Phone: 91-0291-2643993, 2642319
E. mail: agrobiosindia@gmail.com

AGROBIOS (INDIA)

ISBN No.: 978-81-7754-101-4

Published by: Dr. Updesh Purohit for Agrobios (India), Jodhpur
Lasertypeset at: Yashee Computers, Jodhpur
Cover Design by: Reena
Printed at: Agrobios Digitals, Jodhpur

Preface

Experimental embryology is a discipline of developmental biology investigating, problems of differentiation and development of an organism by changing the environment, the cellular composition or the genetic constitution of an embryo.

This book aims to provide the basis of a course in animal developmental biology suitable for advanced undergraduates in biology and for graduates coming, from other fields- assuming no previous knowledge, keeping ideas well afloat without submerging them in a. sea of facts, and covering all of the main aspects of the subject in a way that leads directly into the research literature.

Sufficient descriptive material is woven into the text and illustrations to make the essential features of developmental biology. The main emphasis is upon experimental work which throws light on the underlying mechanisms of development.

Special efforts have been made to compress as much information as possible into the illustrations, which have all been redrawn from various original sources in order to clarify and amplify the text, as well as to convey something of that visual aspect of the subject which has a strong appeal for many developmental biologists and which plays a significant part in research.

I have drawn many simple diagrams from everyday object and discussed new techniques in very simple ways. This manual covers the syllabi for chordate experimental embryology prescribed by various Indian universities including medical institutions.

I am grateful to my colleagues for their many helpful suggestions. Above all, I thank my wife, Shashi for her support throughout the preparation of the manual. Special thanks to Mr. Updesh Purohit, Desk Editor, Agro Botanical Publishers (India) for finalizing the book.

O. P. Jangir

Contents

CHAPTER 1

Introduction

The aim of this manual is to help the students to learn about how animals grow and develop. When you go to college each day, you expect to learn you learn some facts, some ideas, some ways of doing things. You are going to study growth and development not only by reading about them or by being told about them, but by finding out for yourself by actual laboratory investigations. As far as possible, you will conduct a scientific study in the way that a scientist would conduct it.

What Does a Scientist Do?

A scientist who wants to obtain new knowledge by research must begin by asking, himself a question. In our case, the question will be an apparently simple one. What happens as an egg changes into an animal? This question may seem simple, but it is a big question, and a challenging one. Biologists have learned about this tremendous subject, but there is still much to be learned.

As he approaches a new problem, particularly if it involves working with unfamiliar organisms, the biologist begins by observing. He looks at the eggs or the plants or the processes that he is interested in by simple observation aided perhaps by microscopes and calipers and other instruments he learns as much as he can about his material under normal conditions. Keen and thoughtful observation is still one of the biologist's best tools. So we shall begin by looking at frog's eggs and hens' eggs as they develop.

The Experimental Method

As you observe the steps by which the egg changes itself into an embryo and then into a tadpole and than into a frog (Fig. 1) or a chick, you may find yourself involuntarily doing something. You may wonder why some developmental changes go on at a certain rate or what makes some structure appear at a certain time or whether some event that happens early is necessary for another event that happens later. To answer such questions, the biologist resorts to experiments.

A good experiment begins with a precise question that is possible to deal with in the laboratory or in the field. In working out this Block, for example, you will notice that from larvae do a good deal of ineffectual wriggling before they can swim well, and you may wonder whether the wriggling is necessary as practice for swimming. That's the kind of specific question that makes the basis of a good experiment. You will find that you can answer the question by raising the larvae in a substance that prevents the early wriggling movements.

This experiment on the relation of wriggling to swimming is a rather simple one sometimes when you raise a question, you cannot see any obvious way to go about answering the question. In that case, you have to invent a possible solution to the question what the scientist calls a working hypothesis and then try out your invented solution. Take the matter of fertilization as an example, you will see in the course of this Block that frogs' eggs begin to develop when sperms are poured over them. Now you might wonder if the sperms are really necessary, or whether the ripe egg could be made to develop without being penetrated by a sperm. To arrive at a working hypothesis in this case, you might suggest that the essential aspect of fertilization is breaking through the egg surface. Now you have something to test: Prick the egg with a needle and see if it is stimulated to develop.

The Experiment and the Observations

An experiment starts with a question. But the result of the experiment is seldom a "yes" or "no" as an answer to the question. Consider the experiment on the necessity for "practice" of swimming. If the embryos that are immobilized during, development by some chemical substance are, nevertheless, able to swim when they are removed from the substance, then you have shown that practice is not essential. But suppose the tadpoles, when washed free of the substance, are not able to swim this does not prove that practice is necessary. The reason might be that the chemical interfered with the normal development of the muscles, or of the nervous system, so that the tadpole is permanently paralyzed. You would have to deal with these Possibilities by further investigation.

The experiment on pricking unfertilized eggs involves similar difficulties. If the needle-pricked eggs are stimulated to develop normally, you have proved that sperms are not essential for development. A negative result, however, would only show that needle pricks don't cause development. The necessity for the sperm would remain unproved.

In actual scientific research, most of the experiments give results that do not solve a problem, but are only consistent (or inconsistent) with an invented solution to the problem. In a way, `scientists are the detectives who go on piling up circumstantial evidence until they have a very strong conclusion.

An Experiment must have a Control

We said before that a good experiment starts with a very precise question. Another way of saying the same thing is that an experiment should deal with one variable at a time. Suppose you want to find out if frogs' eggs would develop faster at a warmer temperature than ordinarily used, or in salt water than in fresh water. It wouldn't make sense to put some eggs in salt water at a high temperature, and then claim that the result obtained was due to the salt alone. Rather, you would vary one factor in one experiment. You would put some eggs in fresh water in one experiment and put some eggs in fresh water at a high temperature and compare them with similar eggs in fresh water at the normal temperature. Such experiments would enable you to determine clearly the effect of temperature and the effect of salt separately.

In the two experiments just cited, the important point is to make a comparison with the normal situation. The parallel batches of eggs kept under normal conditions in both cases are the controls. In biological experimentation, controls are indispensable. Unless you have a suitable control, you have no way of knowing what your variable factor is really doing.

Because living things tend to differ from each other, the biologist must always have controls and should always use a number of individuals in both his control and his experimental series. Hen's eggs may all look pretty much alike, except for colour, but the embryo inside each egg is an individual, differing to some degree from all other embryos. Consequently, if you want to determine what effect a substance like thiourea has on the growth of the chick embryo, you wouldn't get very far by injecting thiourea into a single egg and then comparing the weight of the embryo at 15 days with the weight of a normal 15-day embryo that you weighed a couple of weeks before. Instead, when you do this experiment, you will have to set up two groups of eggs, treating both identically except that thiourea is given to one group. Then you will compare the average weight of the embryos, in the two groups. If the groups are large enough, the comparison between them will have meaning.

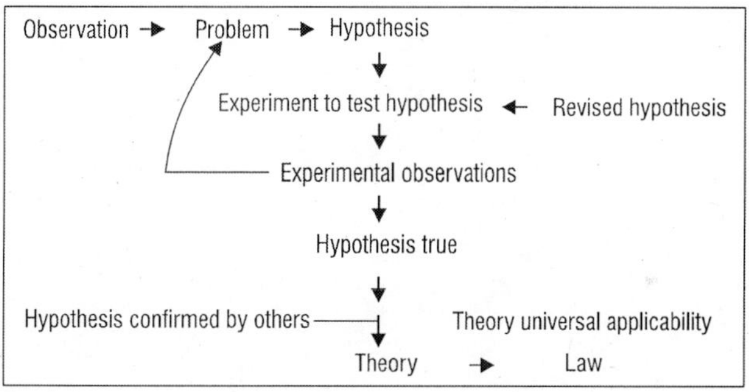

Fig 1.1: Simplified diagram to represent scientific methodology

How to go about it?

Good organization is a key to success in any endeavor. Here are some practical hints that will help you get the most out of this Block:

1. Read each day's assignment in advance and do your best to understand what you are to do. The laboratory periods are too short' for reading the instructions carefully first and then carrying out the procedures. Then, too, you learn more than merely carrying out a procedure when you understand fully what you are doing.
2. Keep a notebook. Every successful researcher has to devise some satisfactory method of keeping his records straight. In the course of this Block you will sometimes be working on two or more different experiments on the same day,

and so you will need to keep a loose-leaf notebook, divided into several sections. Each time you do any job in the laboratory, put down exactly what you did. Be sure to date each entry. Record all observations and weights an,' measurements that you make. Put down questions that occur to you, too. Do not take out pages or improve them because you have made a mistake. Scientists keep a record of everything, including mistakes. Something that seemed quite useless when it was done may turn out to be the most important point. Above all keep your notes up to date, recording each step as you take it.

3. From time to time discuss the work in Block. How useful these discussions will depend on the preparation each student makes. Before the discussion period, think over the questions that have occurred to you consider what the results of your experiments meant see if you can think of other ways to attack the problems of this block. If you and your classmates take these discussions seriously, you may find yourself sharing some of the excitement that fills the air when a group of researchers sit down together to talk about their advances along the frontiers of knowledge.

Note: *Posing a problem or asking a question with reference to the observation is the next step in scientific methodology.*

CHAPTER 2

Collection and Maintenance of Eggs and Embryos

It is suggested that the instructor/teacher demonstrate the following procedure and provide each student with living embryos for study concurrently with the preserved material. Departments which have the facilities may also give the students the opportunity of some simple experiments with the living embryos.

Collecting Eggs in the Field

Spring is the best time for working with amphibian eggs because all but a few species lay in the spring, in ponds, temporary pools and quiet streams. In general, eggs that are laid early in the spring, when the weather may be cold, can endure very low temperatures, whereas those laid later are killed by low temperatures but are able to tolerate quite high temperatures.

Amphibian eggs are generally laid in water. They are surrounded by a clear, soft jelly in which they may float on the surface, or they may be attached to submerged sticks or other objects in the water. In some species, a single clutch (i.e., the eggs produced by one female) will frequently contain more than 2000 eggs. The egg masses may be brought into the laboratory where they may be used for observation or experimental work.

Obtaining Eggs in the Laboratory

To work with frogs' eggs, it is not necessary to depend on collecting them during the short period in the spring when laying occurs. Rather the eggs can be obtained in the laboratory during the greater part of the year. Since the eggs are fully formed months before they are laid, one can induce the eggs to be released from the ovaries by injecting pituitary glands which contains gonadotrophins. This technique is described in the next, chapter.

The frog egg is a spherical (Fig. 2.1) that, in most species, has a darkly pigmented hemisphere and a pale, cream-coloured hemisphere. The creamy material is the yolk, the food store provided for the embryo. When the eggs are stripped from the female and fall on the glass dish that receives them, they are oriented in various ways some with the dark area up, some with the dark area down. When the egg is fertilized, a membrane that was bound tightly to its surface springs away, and the egg is free to move. The pale, yolk-laden area, which is heavier, shifts down-ward so that the dark areas are turned uniformly up. This shift is advantageous for further development because the dark hemisphere absorbs the heat of the sun thus the temperature of the egg is raised a little, and as you will see for yourself in the laboratory, development is speeded up. When you see the dark sides turning uniformly up in the sample of eggs, it means that fertilization has been successfully accomplished.

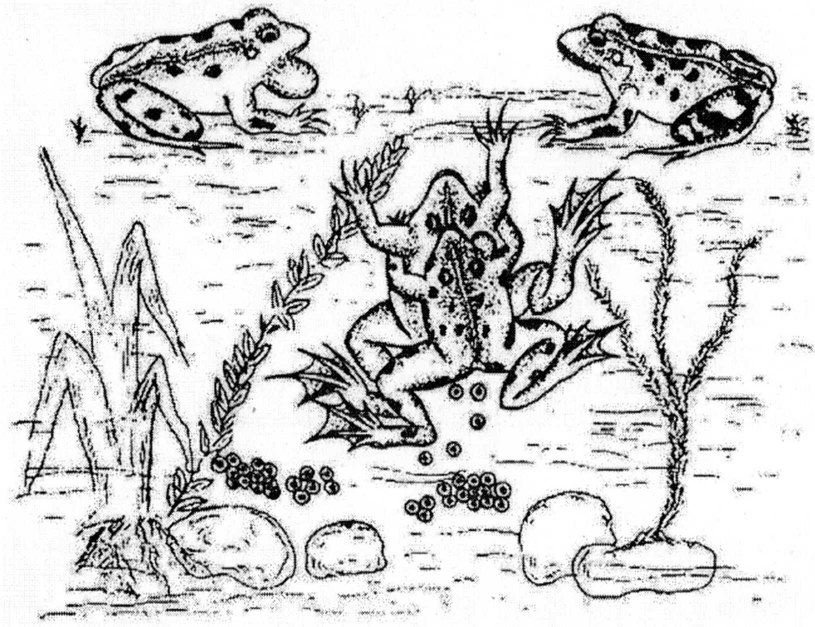

Fig. 2.1: (a) Showing mating call, male, Amplexus in Rana cyanophlyctis and newly laid eggs.

Materials

Insect catching net, Big polythene jars, glass troughs (30 cm diameter, 10 cm deep), Finger-bowls big and small tea strainers, Glass aquaria or enamel pans (30 x 30 x 10 cm), Half boiled spinach or lettuce leaves, Filtered pond water, Petri dishes. Two sharp needles, two pairs of fine watch makers tweezers (forceps) No. 5, Blotting or filter paper, Binocular or dissecting microscope with a light source.

Collection

1. Locate a pond or ditch inhabited by frogs.
2. Collect eggs carefully watch the pond area in the area of vegetation. Clutch of eggs is seen as, a jelly mass. Dip a jar below it and lift it in a container. Alternatively, pick up with a large tea strainer and transfer into a water filled polythene jar.
3. To catch tadpoles, drag the insect catching net in the green vegetation on the side of the pond. Immediately empty the contents in a tray, pick up the tadpoles and transfer in the polythene jar filled with clean pond water.
4. Bring the eggs and tadpoles in well aerated containers, to avoid mortality.

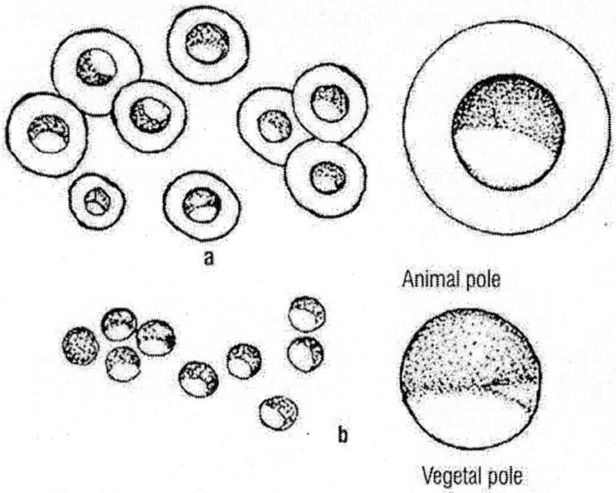

Fig. 2.2: Removal of jelly and egg membrane from the egg. (a) Egg with jelly coat, (b) Egg without jelly Coat

Maintenance

1. For both, the eggs and tadpoles the best medium is clean pond water in which animals normally breed. Maintain 6 tadpoles in 100 ml water in finger-bowls, or 100 tadpoles per aquarium.
2. Feed the tadpoles on half boiled spinach or lettuce leaves.
3. Change the water once or twice in a week. The water should never be allowed to become turbid.

Removal of Jelly and Egg Membranes from the Embryos

To secure eggs for experimental manipulations.

1. Take a few (10-20) eggs in a petri dish containing filtered pond water.
2. Separate jelly from one egg at a time by tearing it off with the help of forceps, cleaning forceps ends occasionally with filter paper. Each egg along with its own jelly measures 3.5-4 mm in diameter. Adjust the mirror in such a way that vitelline membrane shines distinctly.
3. Focus the egg (about 1 mm in diameter) under high power of binoculars to visualize the clear transparent vitelline membrane (0.03 mm in thickness).
4. Insert the needle point in it, pushing the egg slightly to side, with the needle in the right hand, and tear off the membrane by pulling it with the ends of forceps in the left hand.
5. Embryos at the time of release from the vitelline membrane becomes deformed, but regains shape within 5 min. Slight injury is usually unavoidable, particularly in the early stages.
6. Store the embryos in one corner of the petri dish.

Notes

1. For early stage embryos, one must collect before the sunrise, around 3 or 4, since the eggs of Rana cyanophlyctis and other species are fertilized in the morning.
2. Eggs and tadpoles of different species can be identified. On the basis of their body pigmentation, size of, body and way of their swimming etc. The eggs of Bufo form beaded strings whereas Rana eggs are found in clusters.

Solutions

- **Conditioned tap water:** Water left exposed for dechlorination for 24 hours in a vessel with maximum surface exposed to air. Keeping some green plants in it helps the further process.
- **Holtfreter solution:** Weigh the dry salts for appropriate amounts and store in vials. Stock solution is 10x or 2x. Standard solution contains the final concentrations as follows and facilitates wound healing. NaCl 3.5g; KCI 0.05g; $CaCl_2$ 0.1g; $NaHCO_3$ 0.02g; water up to 1,000 ml.
- **Clean pond water:** Allow the water to stand and decant, or filter through filter paper.

Maintaining Living Cultures

Maintaining Tadpoles in the Laboratory

Tadpoles are vegetarians and so are relatively easily kept in the laboratory. They should be kept in an aquarium or a similar container with plenty of water. The water used may be either pond water or tap water which has been allowed to set several days for the chlorine to escape.

Such food as algae, boiled lettuce, or spinach leaves, or any canned, strained, green baby-food vegetables are satisfactory for tadpoles. Algae have an advantage in that they do not mess up the water, and so the water need not be changed for a week or more at a time, but with the other foods, you should change the water after 24 hrs. The tadpoles should be fed three times a week, the amount depending on the size of the tadpoles. Large ones can eat as much as a teaspoon of vegetable matter per feeding.

Maintaining a *Hydra* Culture

Since there may be a number of detrimental factors in tap water, it is desirable to rear hydras in a special Buffer Solution (see in Appendix).

In nature hydras feed on tiny plants and animals. In the laboratory, artemia (or brine shrimp) serve as an excellent food. Since decaying material is harmful to hydras, it is necessary that you drain off the culture solution twenty to thirty minutes after feeding and replace it with new solution.

Hydras will usually stick to the container in which they are being cultured, but some will be poured off with the culture solution. To avoid losing the hydras poured off, pour the solution into a round dish and swirl the dish. The hydras will collect in the center of the dish and may be recovered with a medicine dropper. If the amount of culture solution is so great that this is not practical, you may pass the solution through a 38-mesh net. The artemia will pass through this but the hydras will not. The hydras can then be recovered by inverting the net in the new culture solution.

Culturing Brine Shrimp

Pour 500 ml of artemia hatching brine (500 ml tap water with 2 tablespoons salt) into a large, flat dish and seed it with 3-4 ml of artemia eggs. Do not move the dish after seeding it for you will wash some of the eggs onto the side of the dish where they will not hatch. Cover the dish and let it stand at room temperature. The artemia should hatch in about 48 hours.

To collect the hatched shrimp, place a box with a small hole in it over the dish to darken it. The box should be slightly larger and higher than the dish, there should be a hole about the size of a quarter in the box above the dish. Shine a bright light over the hole and allow about 10 minutes for the shrimp to collect in the lighted area. Draw off shrimp with siphon or large bulb syringe. Empty the shrimp into a 125-mesh net or a piece of cloth, such as a handkerchief, which has been cupped in the mouth of a glass or bottle. Rinse brine from the shrimp with dripping tap water and feed them to your culture immediately by inverting the net or cloth in the culture solution of the animals you are feeding.

Incubation Hen Eggs

Purchase only fertile hen eggs. These may be secured from commercial hatcheries or from poultry farmers who keep roosters with their flocks. Once you have purchased the hen eggs, keep them cool, but not below 10° C, until you are ready to incubate them. The viability of embryos will be greatly reduced in eggs which are kept longer than a week after they are laid.

Hens' eggs are preferably, incubated in a commercial incubator. However you may use any device in which the following conditions can be maintained.

1. The temperature can be maintained constantly between 37°C and 39°C.
2. The humidity can be maintained above 50%. (A large flat pan of water is suitable for this purpose).
3. Ample air space must surround the eggs, so that the developing embryos can obtain the oxygen they need.
4. At least during the first 10 days, the eggs should be turned once each day, otherwise the developing membranes may stick to the shell.

Eggs must be marked with their use and the date they are to be used before placing them in the incubator. Eggs should be marked by writing on the shell with a soft lead pencil.

Culturing Planarians

Planarians can best be kept in pond water in a shallow container and maintained in the dark. They can easily be fed by placing a small piece of raw liver in their culture container and placing the container back in the dark. After about thirty minutes, remove the liver and tap the container lightly on the table. This makes the planarians adhere to the bottom. Carefully pour off and replace the water. Planarians should be fed two or three times a week, and water must be changed after each feeding. Planarians do not need food while regenerating, but you should change the water twice a week. Once a week rinse the dish two or three times by adding water and then pouring it off.

Induced Hibernation in Adult Frogs

Living adult frogs can usually be held over in the laboratory for a week or two after they are delivered by placing them in a moist environment in a refrigerator. This can be done by placing about 6 frogs in a little water (not more than ½ inch deep). The water should be changed at intervals of 3 or 4 days. If any frog dies, it should be removed promptly.

CHAPTER 3

Induced Breeding in *Rana cyanophlyctis*

Frog eggs have been very popular material for embryological studies because of external fertilization, large clutch size and relatively large size of the eggs. In addition, the metamorphosis of the mainly vegetarian aquatic tadpole into the insectivorous amphibious froglet is one of the most fascinating transformations in development.

Ovulation in the mature female frogs occurs only once a year, at some time during the monsoon season. For a while the seasonal breeding of the amphibians proved to be a handicap. However, by induced ovulation it is possible to obtain eggs throughout the year. Recently it has been possible to control' the reproduction also of the Indian skipper frog Rana cyanophlyctis. This enables one to obtain eggs throughout the year.

Materials

Gravid female frogs. Sexually mature male frogs, Dissecting instruments, Stereoscopic binocular microscope with a light source, Petri dishes of 10 cm diameter, Glass aquarium, Filtered pond water, 2 Hypodermic syringes with 18 gauze needle, Glass homogenizer, Troughs to maintain eggs and embryos.

Procedure

Collection of Pituitary Glands (Hypophysis) and Injection in Frogs

1. Hold the frog in your left hand, dorsum up. Insert the pointed edge of a large scissors in the angle of jaw.
2. Cut the head from the body, just behind the tympanic membrane. Discard the body, along with lower jaw, after removing vertebral column to avoid reflex actions of legs.
3. Wash the head. Put upside down under a binocular. Remove the skin of palate to expose the parasphenoid, eyes muscles. Note the location of the hypophysis.
4. Decapitation of frog. Scissors inserted into the angle of jaw.
5. Skin of palate being removed to expose bones and locate pituitary.
6. After removal of bones. Optic chiasma seen as an X, behind this is seen a translucent infundibulum and posterior to it a whitish pituitary body. Black pia mater and medulla oblongata are seen posterior to the pituitary.
7. Insert the blade of a fine pair of scissors in foramen magnum and cut through the floor of cranium on either side, up to the level of eyes.
8. With a pair of forceps lift the cut portion of bone forward to expose the

hypophysis. Locate the hypophysis (Fig.3), it is of different colour than the hypothalamus. (infundibulum). Sometimes it is found adhered to the deflected bone.

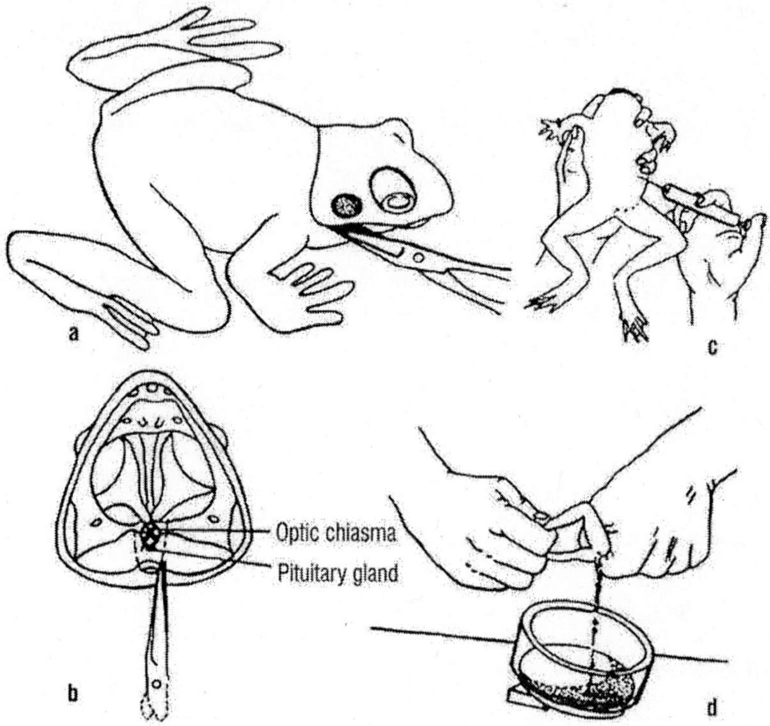

Fig. 3.1: (a) Decapitation of frog, scissor inserted into the angle of jaw. (b) After removal of bones, optic chaisma seen as an X and posterior to infundibulum whitish pituitary body is seen. (c) Injection of pituitary gland. (d) Manner of holding female frog while stripping its eggs into petri dish.

9. Remove the entire gland with the help of forceps and store in distilled water or filtered sterile water.
10. Homogenize as many glands as required in glass homogenizer in such a volume that no frog gets more than 0.5-1.0 ml of homogenate. Each female must receive at least one whole female or two male homoplastic pituitaries.
11. Hold the frog in your left, hand, abdomen up. Inject into the body cavity at the flank taking care that no injury is caused to viscera. Pinch the skin at the site of injection for a while, to avoid extrusion of the injected material.
12. Leave the female frogs in a bowl with water upto 25-50 mm depth.

Preparation of Sperm Suspension

1. Mature sperms are present in the testis of male frog virtually throughout the year although there is a period of relatively lower spermatogenic activity

during winter season.

2. Pith (destroy the central nervous system by inserting a large needle into it) or anaesthetize a male frog.
3. Dissect out the pair of testes.
4. Clean up the adhering blood and mesentery.
5. Cut up the pair of testes into small pieces and prepare sperm suspension in 10 ml of pond water or 12% amphibian Ringer's/Holtfreter's solution by squirting the pieces of testis through a syringe.
6. Allow the milky suspension to stand for 15 minutes during which time the sperms become
7. Place a drop of the suspension on a slide and check for motility.

Ovulation and Fertilization

1. Ovulation ensues within 6-16 hours after injection, depending on the temperature. The time of ovulation sometimes can be determined by the presence of a few eggs in the water. Most of the time, however, the female has to be "stripped".
2. To strip, hold the female between two, hands and apply gentle pressure to the anterior part of the body and then progressively closing the hands toward the cloacal region. If few eggs ooze out of cloaca then proceed to strip.
3. Discard the cloacal fluid along with the initial eggs issued from cloaca.
4. Strip eggs into petri dishes to produce several chains of eggs by moving the female around over the dishes. A mature female of snout-vent length 6.0 cm may release 20-500 eggs.
5. Avoid clumping of eggs.
6. With a dropper squirt the sperm suspension over the eggs,
7. Allow the eggs to remain in the sperm suspension for 15 minutes. Do not allow them to dry at anytime.
8. After 15 minutes period, flood the eggs with pond water (or 10% amphibian saline). Fertilization takes place.
9. After fertilization, the egg becomes activated and rotates slowly within the perivitelline space. Gravity causes the vegetal hemisphere containing relatively heavy yolk to rotate to the underside, with the animal hemisphere (black) rotating to the top.
10. The jelly envelope, initially dense, absorbs water and swells to several times in original thickness. The jelly mass generally sticks to the glass bottom of the petri dishes. Use of scalpel or a blade to free the jelly from the glass and then lift the eggs from the bottom of the dish.
11. With sharp scissors cut the egg mass into small clusters of 15-20 eggs.
12. Transfer the eggs into finger-bowls. For best development keep 15-20 eggs per 20 ml of solution.

Precautions

1. During injection of pituitary extract do not thrust the needle deep into the abdomen. Since such action is likely to stab into the intestine or liver.
2. Check the female after 48 hours by squeezing the abdomen towards the cloaca. If very gentle pressure causes the emission of a few eggs from the cloaca, ovulation has occurred and the frog is ready to be stripped. If she is not ready, put her aside and try again the next day. If the eggs are not needed' immediately after ovulation, put the frog in a little water in the refrigerator and do not keep her for more than 2 or 3 days. Once the eggs have been released from the ovaries, they rapidly lose the ability to survive.
3. When you wish to strip the frog, grasp her in your left hand, as you did for the injection, and hold her legs with your right hand. Then gently press on the abdomen, using your forefinger first, then your middle finger, and so on. Pressure must be exerted toward the cloaca, so that the eggs will move downward and out of the body.
4. The number of glands required to induce ovulation varies with the time of the year and also with the sex of the frog from which the pituitaries were removed. The number of female pituitaries required at different times of the year is included in the following table. If males are used for securing pituitaries, you should double these numbers, since male pituitaries are only about half as potent as those from females.

Number of female frogs' pituitaries needed to induce ovulation at various times

Oct.	Nov.-Dec.	Jan.-Feb.	Mar.-Apr.
8-10	5-6	3-5	1-3

CHAPTER 4

Parthenogenetic Development

Generally the eggs start onset development after fertilization. But it is also possible to make the eggs development without fertilization. The development of the egg without fertilization is called parthenogenesis.

Parthenogenesis occurs naturally among many insects and a few other animals. It is well known that a large proportion, number of the eggs laid by a queen bee are not fertilized by sperm. Instead they are stimulated in some other way they begin developing and become fertile males. The same thing takes place in most of the other social insects, and a few insects reproduce almost exclusively by parthenogenesis. In addition, development without sperm penetration occurs occasionally among many other animals, including vertebrates.

Artificial parthenogenesis may be accomplished in the laboratory with the eggs of many different species. Reproduction of sea urchins and starfishes, for example, normally involves penetration of eggs by sperm. But sea urchin eggs may be activated without sperm entry if they are piked briefly in a slightly acid solution or in a weak solution of potassium chloride or even in diluted sea water. All of these treatments are believed to alter the surface of the egg very slightly, just as the sperm alters the surface in pushing its way through. The result in both cases is that the egg is no longer quiescent but begins to develop. Evidently, ripe eggs contain everything they need to develop, hut are somehow prevented from going ahead on their own. Under normal conditions, it is the stimulus of sperm entrance that releases the egg from its blocked state.

Parthenogenesis has never been observed to occur naturally in frogs or salamanders. But possibly it could be induced in the laboratory by artificial means. Since the frog egg is a rather large, it has been suggested that it might he relatively easy to imitate the penetrating effect of the sperm by pricking the egg with a fine needle point.

As in the experiment "induced ovulation" strip the eggs. These eggs (about five) are taken on a clean slide. Expose the heart of a frog. Puncture the tip of the heart and allow the blood to flow over the eggs taken on the slide. Gently but firmly prick each egg with the sharp point of a glass needle. The prick should be applied within the animal hemisphere (black portion). Immediately after pricking, immerse the eggs in 10%. Holtfreter's solution taken in a petridish. The egg starts development as normal fertilized eggs. Observation is made for every one hour and recorded. Some eggs may develop into tadpoles.

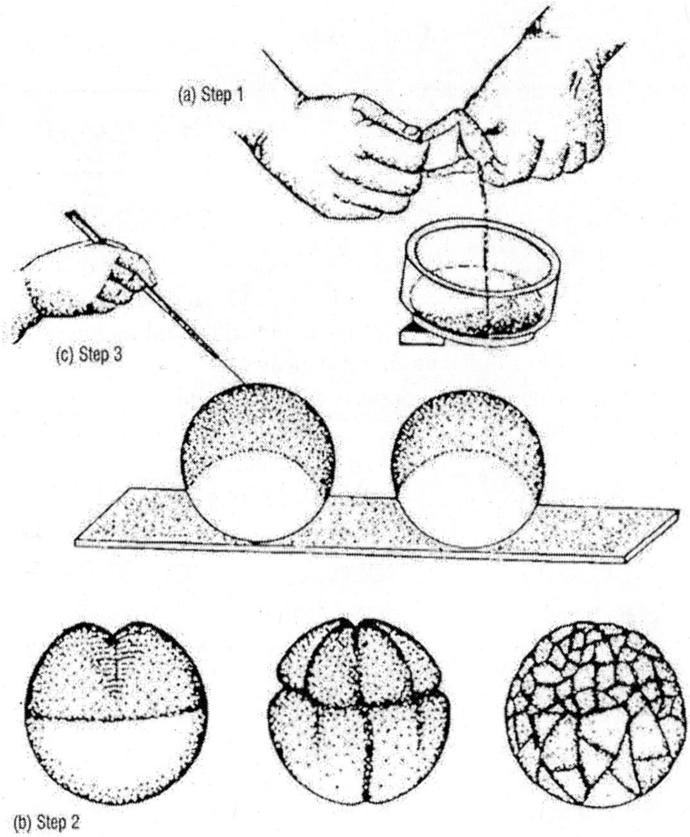

Fig. 4: (a) **Step 1:** Female frog "Stripping" its egg into petri dish. (b) **Step 2:** Eggs on clean slide, allow the blood to flow over the eggs from the heart. **Step 3:** Prick each egg with sharp point of a glass needle within the animal hemisphere. (c) **Step 4:** Immerse the pricked eggs in 10% Holtfreter's solution, Eggs show normal cleavage.

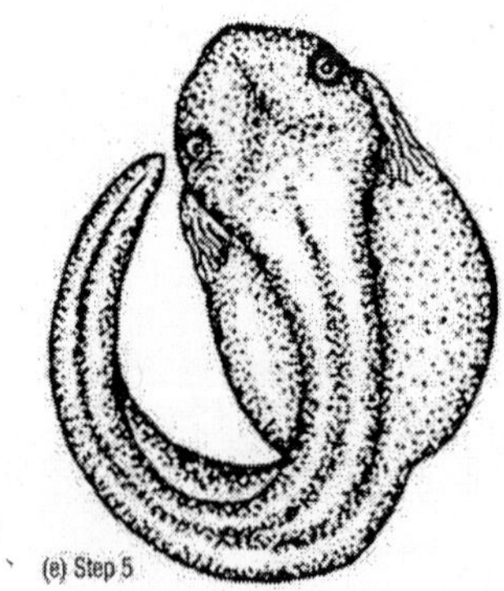

Fig.4.1: Cont. (d) **Step 5:** Parthenogenetically developed tadpole of *Rana cyanophlyctis*

CHAPTER 5

Study of Cleavage Process

Spiral Cleavage and General Development in Snail

The egg cleave in different ways depending primarily on the quantity of yolk contained in them and the genetic determination. Microlecithal and mesolecithal eggs undergo holoblastic cleavage. Eggs of the annelids and molluscs have spiral cleavage while others have radial or bilateral cleavage.

This exercise provides an opportunity to study a cleavage pattern interrelationship between blastomeres and also general developmental processes.

Limnaea, a fresh water pulmonate, is easily available in ponds. Their eggs are collected from the underside' of lotus or water Lilly leaves (Fig. 5. la). The egg masses can be scraped off using a steel spatula. The embryos can be reared in filtered pond water in the laboratory. Individual egg capsules are obtained by rolling the egg mass over a filter paper supported by a glass slide (Fig. 5.1b, C).

Observe the development of Limnaea, eggs from the first cleavage till hatching. After familiarizing yourself with the development prepare a normal table on your own.

Materials

Egg masses of Limnaea, Filtered pond water, 4 cm diameter Petri dishes, Dissection instruments, Binocular microscope with magnification up to 100x. Large troughs of 30 to 50 cm diameter and 10 to 15 cm height.

Procedure

1. Visit ponds with floating lotus or water lily (Nymphaea) leaves. Turn the leaves over and' look for the transparent egg masses attached on the undersurface (Fig.5.1a) to ensure that the snails are breeding in the pond. If so, bring some of these leaves and snails from the pond and put in a trough filled with the filtered pond water. Clean the undersurface of leaves from the attached egg masses, if any, so that the freshly laid egg masses are easily identified.
2. Detach carefully a freshly laid egg mass (each one belongs to a clutch) from the undersurface of the leaf by scrapping by a scalpel or moving one arm of the forceps between the egg mass and the leaf. Transfer it to a petridish containing pond water.
3. Observe under the microscope (30X to 60X) to study the structure of the egg mass and embryo (Fig. 5.1c).

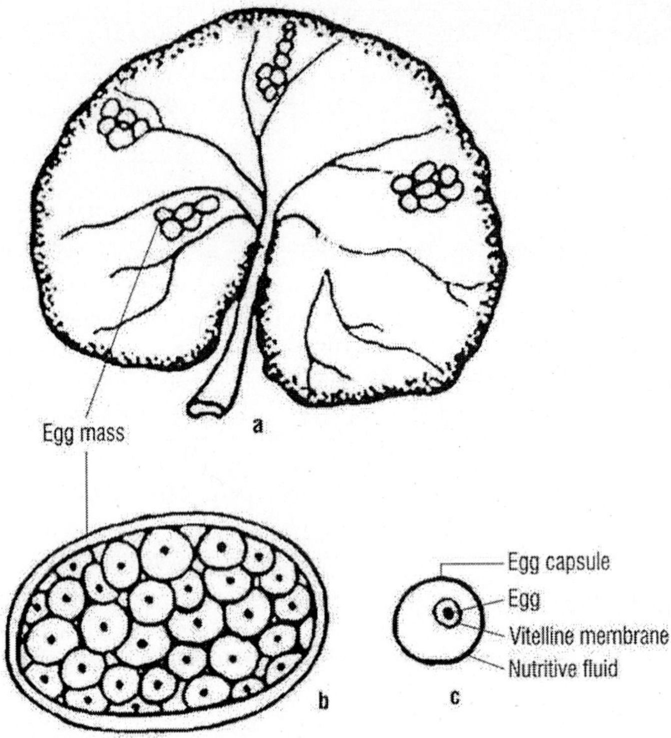

Fig. 5.1: (a) Egg masses of Limnaea attached on ventral surface of water lily leaf. (b) Egg mass in jelly coat. (c) Egg.

Observations

1. Notice the time interval between cleavages. At room temperature it is about 90 minutes. Also note the polar bodies (Fig. 5.2a).
2. Notice the furrow formation, beginning at the animal pole, for the first cleavage (Fig. 5.2b).
3. Notice the flattening of two blastomeres in preparation for the second cleavage (Fig. 5.2c).
4. At 8-cell stage (Fig. 5.2d), notice the relationship between four small micromeres in animal hemisphere, and the four large macromeres in vegetal hemisphere. The cleavage is known as spiral cleavage because the four micromeres chi not sit over the respective macromeres, but on the junctions between macromeres. The direction of coiling of shell is determined by the direction of spiral at the third cleavage.
5. After one day notice of rotatory movements of the embryo organogenesis is achieved. Study the troth, head, foot, visceral mass, shell gland, tentacles, eyes and statocyst (Fig. 5.2e)

6. After 2 days notice the hatching process.
7. Draw the various cleavage stages and later developmental stages, and label.

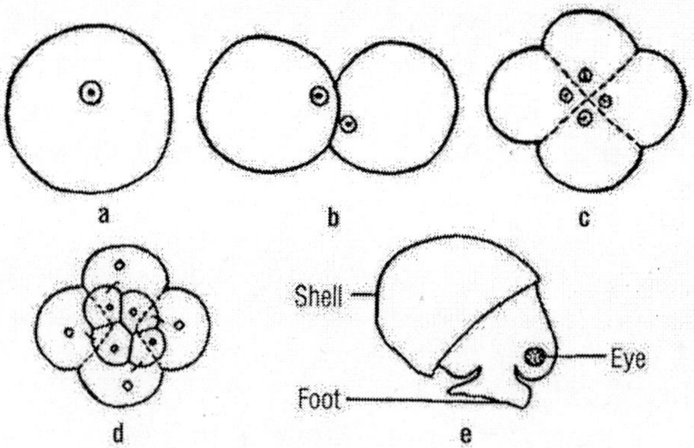

Fig. 5.2: Spiral cleavage in a snail. (a) Fertilized egg. (b) Furrow formation beginning at the animal pole, first spiral cleavage. (c) Two blastomeres, beginning of second cleavage (d) 8 - blastomere stage (e) Newly hatched snail.

Notes

1. Any of the common water snails can be used for the purpose.
2. No aseptic precautions are necessary. Neither the snails need to be especially fed.
3. The experiment can be carried throughout the year but rainy season is best. Snails lay the eggs more frequently in the early morning hours.
4. Experiments like effect of temperature or chemicals on development can be easily carried out.
5. To achieve synchronized and assured egg laying keep the trough free of leaves for a period of one to two days before eggs are required.

CHAPTER 6

Culture of Chick Embryos *In Vitro*

Study of Embryo Recovery and its Culture in Birds

The introduction of more and more pharmaceutical compounds into therapeutics requires in-depth knowledge of their properties. Disasters like those caused by thalidomide emphasized the necessity of testing new substances in order to vouch for their safety. It is of course impossible to test a pharmaceutical preparation for its teratogenic activity in man. On the other hand, it has been demonstrated that animals are more sensitive than human beings.

To screen a substance for its type of toxicity, the growing chick embryo in vitro seems an extremely useful tool it has a convenient size and its organs are easily observable during culture, so that any modification in organ or tissue relations can be followed. Many embryologists proved that culture methods are advisable for explaining early morphogenetic events, and there is no doubt that pharmaceutical and chemical products could be efficiently tested in that manner.

Some of the earliest attempts to culture chick embryos in vitro are as follow:

- **Waddington (1932):** introduced in vitro culture of chick embryos in a watch-glass containing a plasma clot with embryo extract.
- **Spratt (1947):** replaced the plasma clot by a saline agar albumin medium.
- **New (1955):** cultivated embryos on albumen.
- Waddington and Spratt: laid the embryo with its ventral surface next to the medium, while New and Mckenzie laid it on its back.

Spreading the blastoderm on a plasma clot with embryo extract as support and nutrition gave promising results. The watch glass method of growing blastoderms on agar jelly is quite adequate. Most of the carbon and carmine particle marking experiments for tracing morphogenetic movements were done using this technique. A method of growing the chick blastoderm with its vitelline membrane supported on a glass ring was introduced by New. This technique has rendered the older methods almost completely obsolete.

The methods described above are adequate for the study of early developmental stages, generally upto three days of development. Determination of the nutritional requirements, investigating tissue interactions through grafting and chemical manipulation of the embryo's environment are easily achieved by these techniques. Blastoderms of embryos cannot be grown in vitro to stages beyond six days of development. Providing space for the spread of yolk sac and other embryonic membranes becomes virtually impossible. This handicap is overcome by growing the embryos with the yolk balls intact. Whereas the early development is studied more conveniently by isolating and growing the blastoderm, later stages are best studied by the technique of culturing embryos with intact yolk balls. This latter method was developed by Aurbach.

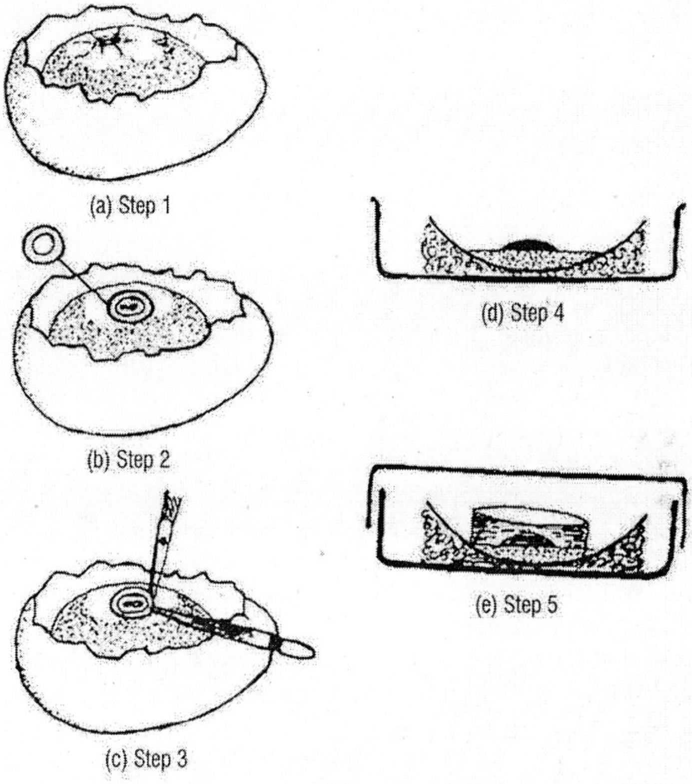

Fig. 6: Culture of chick embryo in Vitro. (a) **Step 1:** Made wide window in the egg. (b) **Step 2:** Place filter paper ring whose inside diameter is barely smaller than the diameter of the marginal blood sinus. (c) **Step 3:** Grasp edge of paper ring and adhering membrane with forceps and clip the membrane completely around the paper ring. (d) **Step 4:** Lift the filter paper ring with adhering membrane and embryo and transfer it on the surface of culture medium. (e) **Step 5:** Cover it by petri dish and put it in the incubator at 37°C.

The chick embryo or its parts can cultured for many days. For the success of the experiment two conditions should.be satisfied:

1. the culture medium should provide all the nutritive components. for the growth and differentiation of the embryo and
2. the medium, the tools, instruments and, glasswares which are used for the experiments should be properly sterilised and free from any type of contamination.

The glassware is thoroughly washed with soap water and rinsed in tap water four or five times. The cleaned glassware is sterilised by keeping it in an autoclave at 100°C for one or two hours. Fresh dissecting instruments should be used. They should be cleaned and kept in the autoclave foe 30 minutes, cooled immersed in 95% alcohol until used. Hands should be cleaned with 95% -alcohol. The table is gleaned and sterilized with 95% alcohol. The cotton, filter paper, etc. should be sterilized properly.

Materials

Watch glass, Petridish, Slides, Cover glass, Pipette, Erlenmeyer, Flask 100 and 500 cc, Finger bowls, Microscope, Filter paper, Fresh eggs, Cotton, Agar (USP X 1). Penicillin, G. Streptomycin, Chick Ringer's solution, Distilled water, Phenol red etc.

Medium

Ringer-albumen, Agar medium, is the proper medium for the culture of chick embryo. This medium is formed of two components. (i) the ringer albumen component and (ii) the Ringer-Agar component. These two components should be mixed only just before use.

For the preparation of **Ringer-albumen component** the white albumen is collected from a fresh unincubated egg. It is taken in a 500 cc flask. To this 50 cc of chick Ringer's solution is added. The flask is stoppered and the contents are vigorously shaked for one minute. Then 0.5 mg of phenol red is added. The phenol red makes the medium slightly purple. This colour forms a suitable background in which the cultured white embryo or its part will be distinct.

For the preparation of Ringer-Agar component about 0.15 gms of powdered Agar (USP X 1) is added to 30 ml of chick Ringer's solution in a small Erlenmeyer flask. This mixture is slowly boiled in a water bath for about 30 minutes. During boiling, the flask should be agitated constantly. This will prevent the sticking on charring of Agar-Boiling helps the agar to dissolve completely. Cool the solution slowly to 45° C. Then 5000 units of Penicillin G, and 10 mg of streptomycin sulfate are added to the Ringer-Agar solution.

Mixing the two components: To the 30 ml of Ringer-Agar component (45°C), 20 ml of Ringer-Albumen component is added. The foamy portion is excluded. The mixture is gently shaked to mix the two components. This is the culture medium. About 2 ml of this medium is taken in the watch glass of each culture assembly. The medium is allowed to gel for about one hour. Now the culture medium is ready for the culture of embryos. A number of such assemblies are prepared and stored in the incubator.

Preparation of culture assembly: A clean watch-glass is placed in the centre of a petridish. The watch glass is kept in position by placing a cotton ring around the periphery of the watch glass. Sterilised water is sprinkled on the cotton ring. The petridish is covered with its cover.

Preparation of the Embryo

Fertile eggs are incubated for about 24 hours. The incubated egg is placed on the table in a horizontal position for two or three minutes. Now the blastoderm moves to the upper surface of the yolk. In the same horizontal position the egg is cracked and the contents are allowed to flow into a finger-bowl containing 100 ml of chick Ringer's solution. Now the embryo floats on the yolk. A filter-paper ring is gently placed on the embryo in such a way that the area pellucida should be seen through the hole. The chalaza is grasped with the help of a forceps held in the left hand. The blastoderm is cut around the periphery of the paper-ring with the help of a sharp scissors held in the right hand. The cutting is made easier by cutting the blastoderm in one direction and pulling the chalaza in the opposite direction.

Now the blastoderm is rolled back and away from the yolk. The yolk sticking to the blastoderm is gently washed away with the help of a pipette. Now the filter-paper ring and the vitelline membrane are carefully removed. Then the blastoderm is transferred with the help of a wide mouthed pipette to a petridish containing about 20 ml of chick Ringer's solution. The yolky opaque area is carefully trimmed with the help of a fine dissecting needle. Now the embryo is ready for cultivation partly or whole.

The embryo is placed carefully on the surface of the culture medium with the dorsal side facing up. This can be done by means of a wide mouthed pipette. The blastoderm is flattened out by sucking away all the excess superficial medium with a fine pipette. The petridish is covered and placed in the incubator. The temperature of the incubator is regulated to 37° C.

Observation

When the embryo is introduced in to the culture medium. It should he carefully examined under the dissection microscope, the various structures of the blastoderm should be recorded on a sketch for future comparison. The observations should he recorded at regular intervals for a few days.

In the above technique, it is evident that:

1. the early blastoderm does not have adequate endogenous food supply but depends upon some exogenous source fir nutrition.
2. in the Ringer-albumen-Agar medium, the albumen acts as a bacteriolytic agent in addition it helps to maintain a proper pH of the medium.
3. the technique also helps to map the various organ-forming areas in a chick blastoderm and it also provides facility for the study of inductive interactions between different tissues, organs and organ systems.

Technique for Removal of the Earliest Stage Chick Embryos

Open the egg by usual technique: At 36 hours or earlier there is little or no red blood, and consequently even the yolk sac is hard to see. On the upper surface of the yolk, however, you will probably see a mottled area, this is the yolk sac, and the embryo is in a somewhat clear area in the center. Drain off excess albumen, as shown in figure, so that the surface of the yolk is not covered by excess fluid.

Now take your agar slide and put 2 or 3 drops of water on it. The dry agar will now swell, and after about 3 minutes you can take your scalpel or razor blade and scrape off a small sheet of agar a squarish piece about 1/4 inch on a side will be excellent. Place this sheet on the spot on the yolk where the embryo seems to by lying. Leave it undisturbed for about 3 minutes, then lift the sheet carefully with your forceps. If the yolk surface was not too wet, the embryo should now be nicely stained, so that the parts of the body can be identified. (If the surface was too wet, the stain has probably diffused around without being taken up specifically by the embryonic tissue).

Examine the preparation with a dissecting microscope or a hand lens. After you have studied the stained embryo in place on the yolk, you may wish to remove it for examining under the compound microscope. You may lift it by the paper ring method. (Fig. 6)

Some other uses of this Technique

An embryologist preparing to operate on a chick embryo begins by sawing open a small square window in the side of the shell. He then stains the embryo and operates on it through the window. When the operation is completed, he closes the window by covering it with a small slip of glass cemented on by paraffin. If sterile conditions have been maintained, the embryo will go on developing and will usually hatch. A variety of operations that have provided evidence bearing on a wide range of problems have been performed in this way.

Embryologists have wondered, for example, whether the thyroid and adrenal glands and the gonads would develop normally if they received no stimulation from the pituitary. This question can be answered in part by staining the 33-hour embryo, by the technique you have used in the laboratory, and then cutting off the anterior end of the head where the pituitary gland would later form. Very fine glass needles are used for the cutting. The embryo's head is abnormal, but the rest of the body develops quite normally, except that it is somewhat small up to the time of hatching. The endocrine glands and other organs can then be examined to see how they have been affected by lack of trophic hormones from the pituitary.

Another question, which has been answered rather simply, concerns the boundary of the brain. At 33 hours the future brain grades imperceptibly into the future spinal cord. Where in later development will one end and the other begin? The answer to this question has come from making small wounds along the neural tube and determining where the resultant scars are located later. This technique has shown that the future brain extends all the way back to the 5th somite.

We are well aware that in higher animals such parts of the body as an eye or a hand will not replace themselves if they are destroyed. (In lower animals, however, regeneration of lost parts is quite common). Does the young embryo have the ability to replace a part that is still in a very primitive stage of development? With the kind of preparation you have made in this exercise, one could rather easily cut off an optic vesicle and determine whether the chick would later have two eyes, or only one.

At slightly later stages embryologists have made extensive use of the operative technique in studying the factors that guide the development of the nervous system. At about 48 hours one can cut off a wing bud or a leg bud. The chick will later lack the part removed (i.e., there is no regeneration in this case). The nervous system can then be studied to find out whether a large limb nerve will develop when there is no limb to be innervated.

These are only a small selection of hundreds of experiments that embryologists have attempted on young chick embryos. These experiments have been of great value in helping us to understand the forces that guide the course of events as a simple egg changes into a complex animal. Such understanding is obviously of tremendous theoretical importance, and like most research, this, work on early embryos promises to be of considerable practical importance too. Already it enables us to comprehend what has gone wrong in the development of malformed babies that are born every year. In the future it may help us find the means to prevent the errors of development that leads to malformation.

Solutions Required

1. **Chick ringer Solution (Locke's Solution):** Sodium chloride 7 gm; Potassium chloride 370 mg; Calcium chloride 180 mg; Glucose 1.0 gm and Distilled water 100 cc.
2. **Frog Ringer Solution (Holtfretes Solution):** Sodium chloride 3.5 gm; Potassium chloride 50 mg; Calcium chloride 100 mg; Sodium bicarbonate 20 mg; Distilled water 1000 cc.
3. **Greacher's Alum Carmine:** Water 100 cc; Aluminium ammonium sulphate 3.0 gm; Carmine 1.0 gm Boil the mixture until the carmine is dissolved. Cool the solution and filter. This stain is used for whole mounts.
4. **Harris Alum Haematoxylin:** Dissolve 1.00 gm of Haematoxylin in 10 ml of absolute ethanol. Similarly dissolve 20 gm of Aluminium ammonium sulphate (alum) in 200 ml of distilled water by heating. Mix the two solutions; boil; add 1.00 gm. of mercuric oxide slowly with constant scarring, boil the mixture again until it becomes dark purple. Cool it by keeping the container in water. After cooling filter it. This stain can be used for sections and whole mounts.

- **Bouin's Fixative:** Picric acid and saturated aqueous solution 75 ml; Formalin 25 ml and Acetic acid (glawal) 5 ml.
- **Eosin (Alcoholic):** Eosin powder 200 mg; Ethanol (70%) 100 ml Dissolve Eosin powder in ethanol. After two hours the stain is filtered through filter paper.
- **Lavdowsky's Fixative:** Formalin 10 ml; Acetic acid (glacial) 2 ml; Alcohol (95%) 50 ml Distilled water - 40 ml.

CHAPTER 7

Study of Chick Embryos by Vital Staining

The vital dye stains the living embryos without killing them.

Requirements

Incubated eggs, microscope, dissection instruments, Locke's solution, alcohol. 1% iodine, slides, cotton etc.

Procedure

The procedure can be divided into following stages:

Preparation of Stain

1. In 50 ml of distilled water 2 gms of Agar is added in a flask.
2. The contents are boiled in a water bath for about 30 minutes. Flask is agitated constantly to prevent sticking or charring of agar during the boiling.
3. On the complete dissolution of agar the flask is taken out.
4. Pour 2 -ml of the content on a microscopic slide with the help of a sterile pipette. Prepare 10 to 20 such slides and dry them for 2 to 3 days in a dust free atmosphere.
5. Dried slides should be stained in 1% neutral red or vile blue sulphate in distilled water for one or two days, Then dry-them again and stored for future use.

Preparation of Embryos

1. Take 24 hrs incubated egg and place on cotton bed in finger bowl. Oriented the egg in a horizontal position with the blunt end towards left of the observer.
2. After four or five minutes the egg is candled against high intensity of light and the position of embryo is marked on the shell with pencil. The egg is replaced in the same position on the cotton bed.
3. The upper surface of the shell including the marked area is rubbed with a cotton swab soaked in 95% alcohol plus 1% iodine. The area is wiped dry with sterile cotton.
4. In the marked area a window is made on the shell. The shape of the window should be square of about 12-20 mm. The window is made with the help of a sharp scalpel. While cutting the shell, the shell membrane should not be damaged.

5. The shell membrane is moistened with one drop of sterile Locke's solution. The window corners are sterilized with 95% alcohol.
6. With the help of a sharp sterile scissors the shell membrane is cut out. The blastoderm should not be touched. If there is excess of albumen it is removed with a sterile pipette.

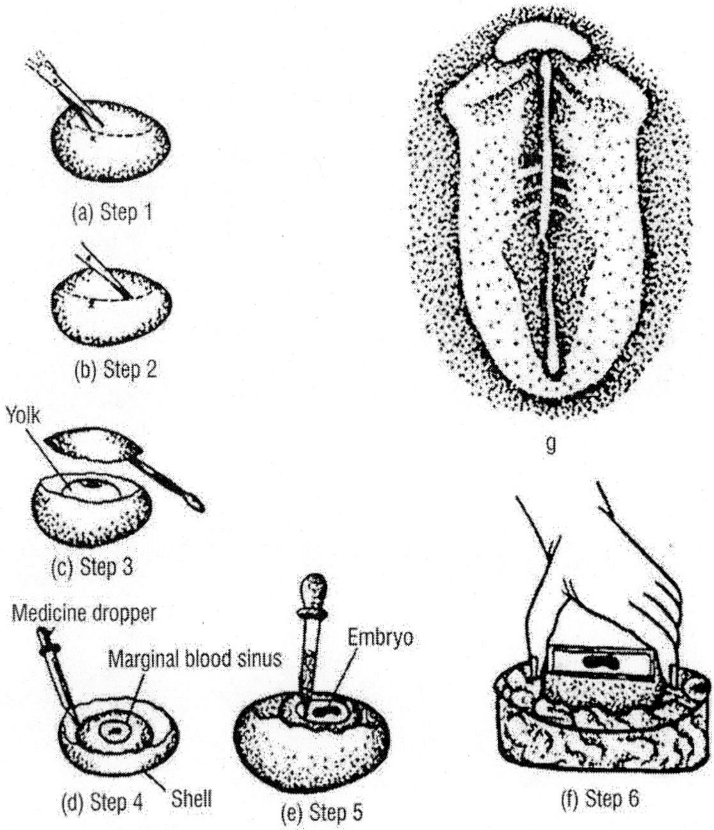

Fig. 7: Study of chick embryo by vital staining. (a) **Step 1:** Insert point of scissor carefully, and barely through the shell at point corresponding to X. (b) **Step 2:** Carefully and slowly clip the shell completely around the egg as indicated. (c) **Step 3:** With forceps carefully lift loose piece of shell and discard it. (d) **Step 4:** Draw off albumen with medicine dropper until yolk surface is not covered. (e) **Step 5:** Place a drop of 0.1% neutral red solution on the aide of embryo. (f) **Step 6:** A coverglass is placed on the window and its edges are sealed with melted paraffin wax. Now the egg is transferred to the incubator and observe the embryo regularly, (g) Chick embryo,

Staining

1. For staining purpose stained agar slide is used. To make this agar moist a few drops of distilled water are added on the slide. The agar swells. Now a piece of agar is cut in a required size. It is placed on the centre of the blastoderm

with the help of a sterile forceps, or place a drop of 0.1 % neutral red solution on the blastodisc.

2. The agar is removed after a few minutes.
3. The embryo is observed and sketched.
4. A cover glass is placed on the window and its edges are sealed with melted paraffin wax.
5. Now the egg is transferred to the incubator and the embryo should be observed regularly every 6 or 12 hours.

CHAPTER 8

The Technique for the Whole Mount Preparation of Chick Embryos

The egg of the hen is a polylecithal type. It develops only on a limited range of temperature (about 37°-39°C), which is approximately the body temperature of the adult bird, therefore the temperature of incubator must be carefully controlled.

Materials

Fertilized eggs incubated for 16 to 18 hours at 38°C, Laid hen eggs can be stored between 7 and 15°C, in a refrigerator, for at least a week, without a noticeable increase in abnormal development. 0.9% Saline, 0.1% Neutral red solution (prepared in distilled water). Dissecting instruments, Finger-bowl of 200 ml capacity and Binoculars.

Procedure

1. Observe the egg. The shell is oval in shape, one end is tapering and one end is rounded. The tapering end is called narrow end and rounded one as the broad end.
2. Crack a small area of egg shell at the broad end with a blunt forceps or with the handle of scalpel (Fig 8a). Remove the shell pieces gently, so that one can see white opaque shell membrane (Fig 8b). There are two shell membranes. The two membranes are closely attached to each other except at the blunt end where they separate from each other, to enclose a small gap in which the air is trapped. Because of this gap at the blunt end, it is easy to open the egg from the broad end without disturbing the embryo.
3. Puncture the inner shell membrane. See yellow yolk ball, with the blastodisc, floating in the thin albumen.
4. Remove albumen as much as possible, with a pipette.
5. Broaden the shell opening (Fig. 8c) so that you can easily pour the entire contents of the egg in the finger-bowl (Fig. 8e). Rotate the yolk ball gently and let the blastodisc be towards the observer.
6. Next step is to cut a filter paper ring with an inside diameter slightly smaller than that of the marginal blood vessel that encircles the vascular part of the yolk sac. Then be sure that the surface of the yolk where the embryo is lying above the level of the surrounding fluid.
7. Now place the filter paper ring around the embryo (Fig. 8d).

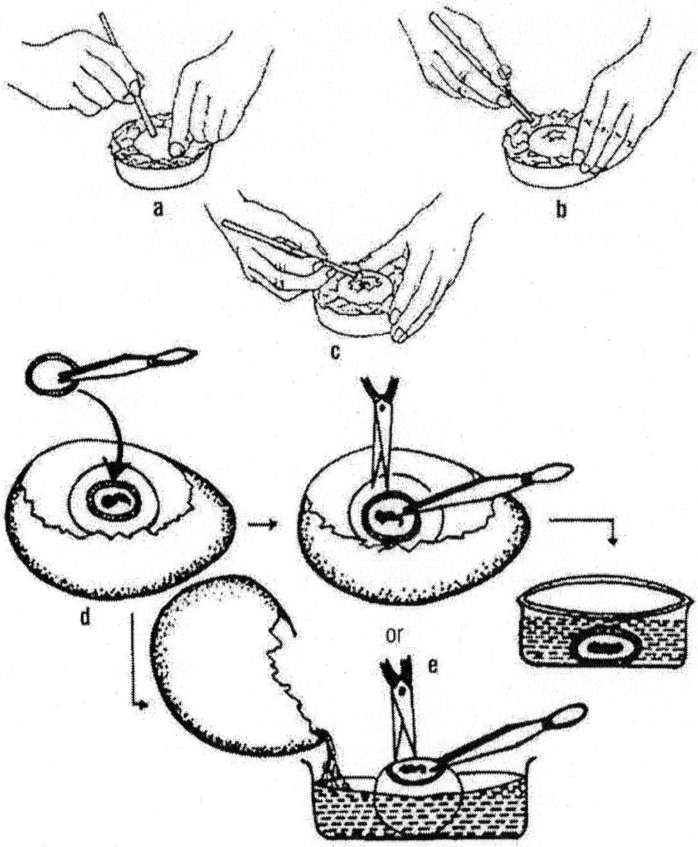

Fig. 8: The technique for the whole mount preparation of chick embryos, (a) **Step 1:** Crack the egg with the handle of a scalpel. (b) **Step 2:** Using forceps carefully pick off bit of shell. (c) **Step 3:** Pick off bits of shell to make window in the egg. (d) **Step 4:** Place filter paper ring whose inside diameter is barely smaller than the diameter of the marginal blood sinus. (e) **Step 5:** Grasp edge of paper ring and adhering membrane with forceps cut round the ring and lift the embryo along with paper ring.

8. Proceed to remove the embryo and the surrounding membranes from the yolk.
9. The embryo is placed in a small petridish in 3-4 ml of warm saline.
10. It is important that the preparations be kept warm all the times. The embryo will live for hours if the temperature is kept near 38°C.
11. Draw off the cool saline with a medicine dropper and replace it with fresh warm solution.
12. Study the embryo under microscope.
13. Carefully remove the vitelline membrane.
14. Pour few drops of Bouin's fixative or buffered formalin solution.
15. After 24 hrs rinse the embryo with water several times to remove fixative.

16. Pass the embryo through various steps of dehydration upto 70% alcohol to which a small amount (2%) of NH_4OH has been added. This will remove all of the yellow stain and then stain Time for staining:
    1. 24 hours .... 1½ minutes
    2. 48 hours .... 3 minutes
    3. 72 hours .... 5 minutes
17. Older stages may be over-stained for several hours and then later de-stained to the proper degree in acidified 70% alcohol. However, whenever de-staining by acidification, there must follow a neutralizing by thorough washing in pure 70% alcohol or better the addition of a slight amount of ammonia (NH_4OH) in the 70% alcohol, with Azocarmine. Dehydrate in 90% and absolute alcohol and then in xylol and mount in DPX or after embryos are fixed they are washed twice with water, processed (first dehydration with alcohol then xylene followed by rehydration, staining with hematoxylin, and then again dehydration) and mount in DPX mountant for morphological study.

Observations

1. A thick covering of albumen is present around the yolk ball (Fig. 8d). Part of this albumen is very viscous and is called, thick albumen. The free flowing part is called thin albumen and can be easily poured out while removing the shell. On either side of yolk a ball is twisted, white cord-like structure is seen. It is called chalaza. It holds the yolk ball in place. It has formed due to the coiling of the thick albumen when the egg rotated during its passage down the oviduct.
2. When the albumen is completely removed, a thin semi-transparent membrane can be seen covering the entire yolk-ball. This is called the vitelline membrane.
3. The blastodisc is a small, white, circular disc on the yolk ball. Its central transparent region the area pellucida is somewhat pear-shaped and the outer opaque region is the area opaca. In the centre of area pellucida 2/3 of it, is covered by a thin primitive streak. The length of the streak depends on the stage of development.
4. Late staged embryos show mesodermal somites and several other structures. Observe the embryos as whole mount under binocular microscope. Note down various structural components of the embryo. Draw a neat and labelled diagram of the preparation.
5. Temporary mounting of chick blastoderm. Remove the blastoderm to a watch glass as described in the previous experiment. Drain off the medium. Add one drop of the stain for a minute. Then drain off the stain. Wash the blastoderm with fresh saline solution. Then, mount the blastoderm on a glass slide with glycerine. Examine under the microscope.

Permanent Mounting of Chick Embryos

1. Shell is cracked
2. Yolk ball is poured in to the warm water,
3. Rotate the yolk ball in such a way so that blastodisc should be on the upper surface of yolk.
4. Put the filter paper ring around the embryo.
5. Blastoderm is cut around the periphery of paper ring.
6. Remove the embryo with paper ring and study in living condition. Then put in watch glass,
7. Solution of watch-glass is replaced by fresh water and then Bouin's solution.
8. Filter paper ring and vitelline membrane are carefully removed.
9. Decolouration is done by adding 3% by volume of NH_4OH to 70% alcohol.
10. Hydrate the fixed embryo by adding distilled water for 10 to 15 minutes.
11. Stained the embryo by eosin for 3-15 minutes.
12. Dehydration by alcohol series 30% → 50% → 70% → Eosin → 90% → 100% (10-15 minutes in each change).
13. Cleaning in 100% alcohol + Xylene (for 10 minutes).
14. Xylene for 10-15 minutes.
15. Mount in canadabalsam.

CHAPTER 9

Study of Mitosis

To study the Chromosome Preparation and Mitotic Index in Regenerating Limb/Tail Lip of Amphibian Tadpoles

Chromosomes are the carriers of genetic material. A study of chromosome complements offers many uses, like sex determination in early developmental stages, detection of variation in chromosome number etc. In present experiment amphibian tadpoles are being used as the experimental material. Amphibian tadpoles have good capacity to regenerate the amputated limb. The regenerating blastema and tail of developing tadpoles are growing very rapidly in the initial stages of development. So they provide an ideal material for obtaining dividing cells. These cells, can be arrested at metaphase, to obtain clear chromosome spreads to study the number, size of chromosomes and mitotic index etc.

Materials

Tadpoles with blastemas on their amputated limbs and those with growing tail. Tap water or pond water, colchicine solution 100 µg/ml. Acetocarmine 40% acetic acid. Petri dishes or finger bowls; slides; coverslips. Paraffin wax and brush. Dissecting instruments & microscope.

Procedure

1. Select healthy looking, developing tadpoles.
2. Amputate the limbs through shank in 10-15 tadpoles and rear them in water. After 2-3 days of amputation blastema is formed.
3. Select 10-15 another healthy developing tadpoles for study of mitotic index in their tail tip.
4. In a finger bowl prepare about 100 ml of Colchicine solution of 10 lag/ml concentration using stock colchicine solution.
5. Transfer 10-15 tadpoles from both groups to the working Colchicine solution and leave them there for about 2-3 hrs.
6. Take a treated tadpole on slide and cut off 2-3 mm long end part of the regenerating blastema and the tail using a sharp blade and needle.
7. Transfer the cut piece to a drop of acetocarmine on a clean slide.
8. Stain for 15-20 minutes. During the staining care should be taken not to allow the tissue to dry. Fresh drops of stain should be added if necessary.
9. Wash the excess stain by 40% acetic acid. This should be done quickly so as to remove only superficial stain and not to de-stain the cells.
10. Place a fresh drop of 40% acetic acid and put a coverslip.

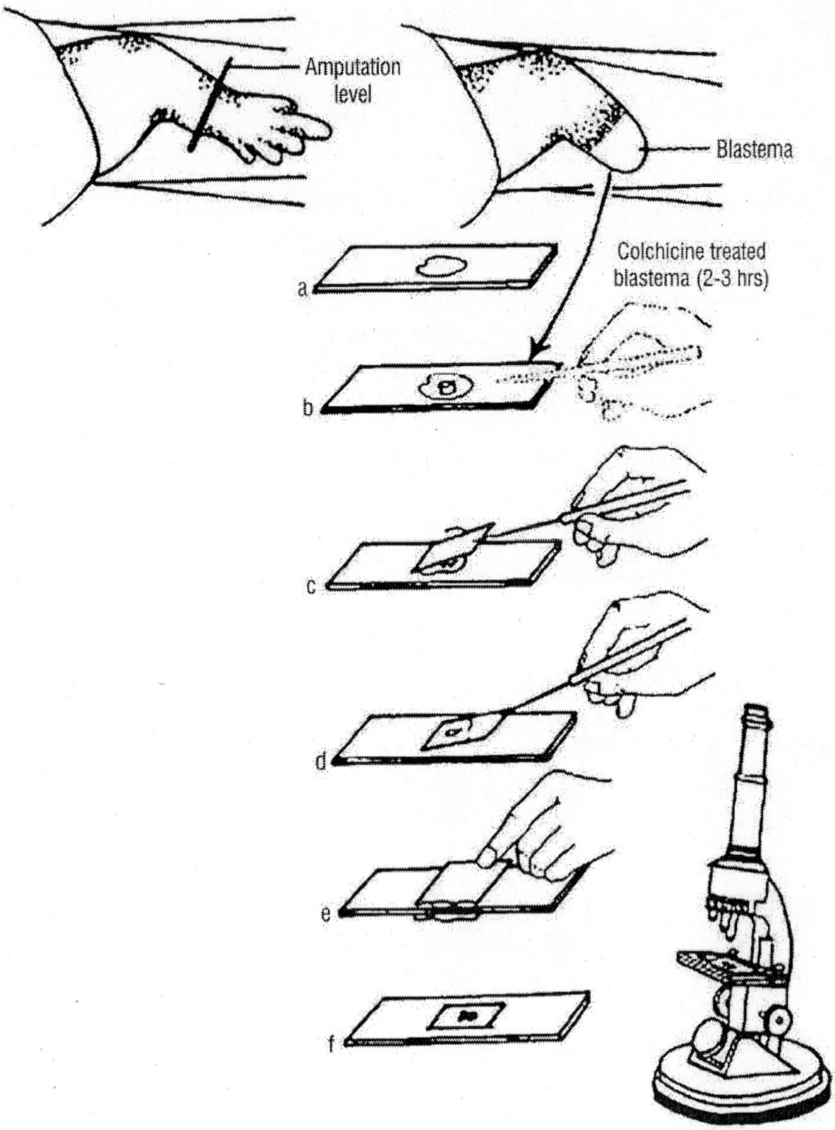

Fig. 9: Study of mitosis. (a) Slide with drop of acetocarmine (b) Transfer the blastema to the slide and stain for 15-20 minutes, then place a fresh drop of 40% acetic acid. (C) Touch cover glass on solution and support by needle. (d) Lower cover glass on to the blastema. (e) Warp a piece of blotting paper or filter paper round the slide and then press cover glass to squash the blastema to obtain a single layer of cell. (f) Transfer slide to microscope.

11. Squash the blastema/tail tip to obtain a single layer of cells.
12. Scan the slide for proper spreading of the cells observe the dividing cells,

most of which will be in a metaphase.

13. Apply paraffin wax to the edge of the coverslip to prevent drying of preparation.
14. Store at 4°C necessary.

Observations

1. Select a slide with good chromosome condensation of metaphase stage and without overlap of chromosomes.
2. Count the number of Chromosomes in several metaphase spreads to determine the number of Chromosomes for the species.
3. Determine the number of cells undergoing division to calculate the mitotic index
4. Mitotic index is number of cells undergoing mitosis per thousand cells.
5. In present experiment histological serial sections of 1, 2 and 3 days old regenerates were made. Average numbers of dividing cells per section of regenerates of different ages were calculated.
6. In the blastemal mesenchyme of the regenerates few cells were found dividing one da] after amputation (2.70.152). In 2 days old regenerates a large number of blastema were found in mitosis (41.1 1.724) and in the regenerates sectioned 3 days after amputation also many cells were in mitosis but the frequency of cell division had come dowi (19.30.445).

Note: *The frequency of cell division may vary according to species and stage of tadpoles.*

Techniques for Chromosome Preparation

Chromosome Preparations of Bone Marrow Cells

Studies of mammalian chromosomes have advanced rapidly through the introduction of preliminary treatment of colchicine and hypotonic solution especially when used for cultured tissues and cells° from peripheral blood. Studies of chromosomes from bone marrow cells, either from short-term culture or directly after aspiration, have aided progress. We wish to report two methods for making chromosome preparation directly from Bone Marrow cells. The first, similar to that used for chromosomes in meristematic tissue to plants permits preparations to be made without prior in vitro culture or in vivo colchicine administration, but requires squashing. This method is especially suited for use where laboratory facilities are limited. The second method, dispenses with squashing but requires the use of centrifuge. In this second method, the cells are fixed and then attached to slides by air drying method. This minimizes mechanical distortion and prevents ruptures, and ensures uniform flattening of the cells in division. These preparations are especially suitable for analysis from photomicrographs.

Requirements

Rat, 0.2% colchicine, syringe, hypotonic solution, test tubes, fixative, stain, alcohol grades, xylene, D.P.X.

Procedure

1. Inject colchicine into the rat (0.2 ml/100 gm) and leave for 2-2½ hours.
2. Aspirate femoral marrow and keep it in hypotonic solution for 20-22 minutes at 37°C.
3. Then centrifuge at 500 rpm for 5 minutes. Remove supernatant and add 2-3 ml of fixative. Shake to loosen cells and leave for 30 minutes.
4. Centrifuge and remove the supernatant. Add 5 ml of fixative (methanol acetic acid, 3:1), resuspend by shaking and leave for 2-5 minutes. Repeat this procedure-once again.
5. Centrifuge and remove all the supernatant, add few drops of fresh fixative and agitate to resuspend the cells.
6. With a long nozzled pipette put 1-2 drops of the cell suspension on chemically clean slides. Blow on each droplet as soon as it is in place to assist in spreading and drying. Leave to dry thoroughly.
7. Stain the slides in 2% Giemsa.
8. Wash in distilled water and mount in D.P.X.

Observations

Prominently stained nuclei can be observed. On close examination metaphase plates can be identified.

Preparation of Meiotic Chromosome from Grasshopper Testes

Materials

Male grasshopper, insect saline (0.67% NaCl), Farmer's solution (1:3 acetomethanol), 2% acetocarmine, 2% acetoorcein, 60% acetic acid. D.P.X.

Procedure

Squash Preparation

1. Dissect out testes from male grasshopper in insect saline solution.
2. Clear out the fat bodies from the testes follicles.
3. Remove the saline solution and o.dd Farmer's solution. (Fixative). Fix for about 15 minutes with two changes.
4. Remove the fixative and add 2% acetoorcein and 2% acetocarmine (1:1) and stain for 15 minutes.
5. Wash with 60% acetic acid. Take one or two follicles on a clean slide, cover with a coverslip and squash gently.

6. Seal it with sealing wax.

Air Dry Technique

1. Proceed till step 4 as above.
2. Wash with 60% acetic acid.
3. Take a small drop of 60% acetic acid on a clean slide and put a testes follicle on it and leave it till it dissolves.
4. After the follicle has dissolved, add a little acetomethanol (1:3) and dry it with hair drier.
5. On complete drying, stain with carbol fuchsin.
6. Mount with D.P.X.

Observations

The cells of testes are undergoing spermatogenesis which involves the process of meiosis, Meiotic divisions are: Prophase I includes leptotene, zygotene, pachytene, diplotene, and diakinesis. Other stages of meiosis I are Metaphase I, Anaphase I, followed by second meiotic division i.e. Prophase II, Metaphase II, Anaphase II and Telophase II.

Study of Cell Cycle

The period, from the formation of a cell to its division into two daughter cells is known as the cell cycle. The cell cycle has three main aspects: (1) Growth, (2) Replication, (3) Distribution.

Growth and replication occur during an interphase and distribution as well as separation occur during actual division (mitosis). However cell cycle consists of periodicity of certain molecular events and hence is divided in five phases: GO, G 1, S, G_2 and M.

Phases of Cell Cycle

Name of Phase	Principle cell activity
GO	Decision as to whether cell should divide or differentiate is taken - Trigger stage
G_1	Initial growth following division - Growth in quantity of cytoplasmic matter
S	Replication of DNA, histones and chromatin
G_2	Continued growth and preparation for division Microtubules formation
M	Division of nucleus, separation and distribution of chromosomes and cleavage of entire cell into two cells.

Autoradiography

Cell is a very dynamic unit of life. A valuable technique for tracing such dynamic events in cells is autoradiography. It is useful in localizing specific molecules, in tracing the pathways of their synthesis, in studying their movements within the cellular compartments and also in studying the fate and breakdown of the molecules. In fact it is the only technique which can help us study the ever-changing aspects of cells. In this technique radioactive isotopes are incorporated into precursors which are used in the construction of specific molecules. For example radioactive amino acids are used to localize or trace proteins; radioactive thymidine is used to detect or to study synthesis of DNA; radioactive uridine is used to study RNA. The precursors are made radioactive by replacement of a normal atom by a radioactive atom such al P_{32}, S_{35}, C_{14} or H_3 in their structure. When the radioactive precursor is taken by the cell, ii will be used in the synthesis of larger molecule.

The technique of autoradiography is based on the fact that radioactive decay expose, photographic film just as light does. Therefore, by placing a photographic film or emulsion over the cells of tissue which have incorporated radioactive precursors and leaving it in dart for several days, the radioactive decays emitted from the atoms of the molecules will expos (the photographic layer. When subjected to standard photographic developing, the exposed site appear as dots or grains over the cell or tissue region containing the radioactive atoms Thus, it is possible to visualize or localize specific molecules. The radioactive decays in the form of B-particles strike silver salt of the photographic film and when the film is developed i results into metallic silver which appears as a dark dot or grain. The grains produced in the film or emulsion are similar, irrespective of the precursor but their size depends on the quality of the photographic material. Autoradiography can also be applied at the ultrastructural level using transmission electron microscope. No. of grains are proportional to the rate and amount of synthesis of specific molecule. Quantitative autoradiography has been useful in ou understanding of synthesis and movements of molecules like RNA from (nucleus to cytoplasm) one compartment to the other.

Methods

Routine Preparation of Tissues for Autoradiography with Tritiated Thymidine

1. Fix tissues in 10% cold buffered formalin or other appropriate fixative.
2. Wash in running H_2O to eliminate chemo graphic effects on film from fixative.
3. Dehydrate and embed in usual way.
4. Cut sections of 5 microns or less.
5. De-paraffinize in xylol and alcohols, gradually descending to 1120. Wash for minimum 15 min.
6. Bring emulsion (NTB_2 or G-5) to room temperature while de-paraffinizing slides.
7. Place in water bath at 43°C for approx. 15 min. to melt emulsion. Rotate bottle gently and dilute amount needed with an equal part of dist. I-120. Mix gently and pour about 40 ml into Coplin Jar which has been warmed to 43°C. Mixture will come to required temperature in another 5-10 min. and dipping may be started.

8. Dip de-paraffinized slides and place in rack to dry (slides should not be too wet before dipping).
9. Store in light-tight slide box containing Drierite and seal with tape.
10. Expose for appropriate length of time in refrigerator set aside for this purpose.
11. Bring to room temperature before developing slides.
12. Develop 2 minute in D-19, rinse, then 2 minutes in Hypo,
13. Wash in running water for 15 min. and apply staining procedure desired.
14. For routine H & E, place slides in Harris hematoxylin for 2 minutes (depending on strength of hematoxylin).
15. Wash in running water for 30 minutes minimum. (If stripping film was used instead of emulsion stain at least 5 minutes in hematoxylin).
16. Counterstain in eosin for 3 seconds (if quick dips). (Stripping film counterstain 2 quick dips with eosin-preferably no eosin should be used- it makes the background too dark and too difficult to see the cells.
17. Eosin used- 1% aqueous eosin Y-50 cc, 80% alcohol-150 cc, Glacial acetic acid-1 cc.
18. Rinse in water and dehydrate in ascending alcohols.

    70%.......... 30 sec.
    80%.......... 30 sec.
    95%.......... 30 sec.
    95%.......... 1 min.
    4 absolute alcohols.......... 3 minutes
    4 xylol.......... 3 minutes

19. Mount with DPX.

***Note:** 500 slides can be dipped with one 4 oz. bottle of NTB.*

***Emulsions:** NTB- Kodak, Stripping Film: AR-2-Kodak, G5-Ilford.*

Autoradiographs of Leucocyte Blood Cells for Chromosomes

1. H^3-thymidine treated blood cells are prepared according to the air-dry technique of Moorhead-Exptl. Cell. Res. 20-613-1960.
2. Chromosomes are stained by carbol fuchsin (Carr and Walker-stain. Tech. 36-233-1961).
3. Differentiate in 100% alcohol -2 changes.
4. Let slides dry in air.
5. Before dipping in emulsion, slide is immersed in warm water for few min. (40°C).
6. Prepare autoradiographs as mentioned above following steps 6 to 13. See A-Routing Preparation.
7. Pass the slides in ascending grades of alcohols.

8. Dip in xylol and mount in DPX using large rectangular coverslips.
9. Study the pattern of label on chromosomes.

Carbol-Fuchsin STAIN

Soln. A.	Basic Fuchsin	3 gm.
	70% Alcohol	100 ml.
Soln. B	Solution A	10 ml.
	5% Phenol sol.	90 ml.
Staining Solution	Soln. B	45 ml.
	Acetic Acid	6 ml.
	37% Formaldehyde	6 ml.

Autoradiography for Studying Cell Cycles of TC Cells

1. Take any healthy rapidly growing culture-line e.g. HeLa, CHO, Fibroblast-L Cells etc.
2. The culture should be in monolayer preferably in Leighton tubes growing on long coverglasses (10 X 50 No. 1).
3. Treat the cells in previously used but good growth medium containing H_3-thymidine (1.0 u Ci/ml-H_3-thymidine Sp. Act. 3 to 8 Ci (m Mole) for about 30 minutes. Change the medium with some non-radioactive (cold) thymidine (1 µg/ml).
4. Remove samples every hour for 20 hours and fix the cells in acetomethanol (1:3).
5. the coverslips with fixed cells are rinsed thoroughly.
6. Autoradiographs made as per steps 6 to 13 described earlier in routine preparation (A).
7. Stain with 2% Giemsa in Sorensen's buffer (pH 7.0) for 10 min. and rinse.
8. Pass through ascending alcohol grades.
9. Immerse in Xylol and mount in DPX.
10. Score under microscope labelled metaphases from samples collected at interval of 1 hr. for 20 hours.
11. Prepare a graph.

$$\text{Total cell cycle} = G_1 + S + G_2$$

$$G_1 = \text{Cell cycle (Hrs)} - (S + G_2 \text{ hrs.})$$

Thus we can get duration of all the phases of the Cell Cycle.

Squash Technique

1. For root-tips prepare squash as usual. The root-tips must be previously exposed to radioactive precursor e.g. 113-Thymidine for two hours before introducing colchicine.
2. Place the slides on a block of dry ice (Solid CO_2) for more than a minute.

3. Remove coverslip by lifting with a razor blade and discard it. Fix in 95% alcohol immediately.
4. Slides are again passed through descending grades of alcohol 70%, 50% and 30% and to distilled water.
5. Follow steps of coating emulsion as in (A) 6 to 13.
6. Pass through ascending grades of alcohol and mount in DPX.
7. Observe under microscope labels (grains) on chromosomes.

Development, Fixing and Washing Autoradiographs

1. Strength D19 at 17-18°C for 4 min. in complete dark. Shake gently while developing.
2. Rinse in water 30 sec.
3. Strength Kodak acid fixer 10 min.
4. Wash 5 min. in water at 18°C. Change water, 2 or 3 times in this period.
5. Strength Kodak hypo-clearing agent (dilute 1:9) of the concentrated solution in the Lab. Use for 1-2 min.
6. Wash 5 min. with 2 or 3 complete changes of water.
7. Dry the slides or stain if necessary.
8. Stain for contrast by transferring slides to solution as follows:

10 mg. azure B Bromide, 500 mg. Potassium acid phthalate, 50 ml. of water.

↓

Stain for 20-30 min. at 18°C.

↓

Rinse ½-1 mini in water, drain off water and dry.

↓

Dry slides in the drying box. Slides can be stored dry without cover glasses

For Removal of Silvergrains after Study

1. Put slides in dist. water for 10 min.
2. Remove silver grains with 7.5% soln. of $K_3\,Fe(CN)_6$ for 3 min.
3. Transfer to a 20% sol. of $Na_2S_2O_{35}H_2O$ for 3-5 min.
4. Rinse in water: 3 changes of 1 min. each.
5. Immerse in 0.1% trypsin at 37°C for 10 min.

   Tris 0.2 M at pH 7.0 .... 50ml.
   Dist. Water .... 50 ml.
   Trypsin (Difco) .... 0.1 gm.

6. Wash in tap water for 5 min. and rinse in dist. water.
7. Dry in air.
8. Transfer to Xylol. Study with oil.

CHAPTER 10

Demonstration of Cell Death

The aim of this experiment is to observe cell death during embryonic development in chick and to assess the role of cell death in morphogenesis and histogenesis.

Cell death is a normal phenomenon during morphogenesis and histogenesis of organs and tissues. It appears to be genetically programmed. It possibly serves as an important mechanism for integration of cells into tissues and organs, by removing the superfluous cells. It is also probable that the necrotic cells release specific signals which may direct the neighboring cells to choose a particular developmental pathway.

The vital dyes are the dyes that are taken up by the dead cells. The phagocytosed particles of the dye are accumulated in the lysosomes. The lysosomes, subsequently phagosomes, are most abundant in the macrophages. The macrophages, having accumulated large quantities of the vital dye inside, appear as small dots under the low power of magnification (x 30). A concentration of macrophages in an area is indicative of the presence of necrotic cells in the area because macrophages accumulate in the vicinity of dying cells to engulf them.

Materials

Embryos chick from limb bud stage to digit formation stage. Physiological saline (0.9% NaCl for birds and mammals) 10-3 Dilution neutral red solution in physiological saline. Watch glass/cavity blocks. 4 cm Diameter petri dishes. Wide mouth Pasteur pipettes. Dissection instruments. Binoculars with magnifications up to x 45 Incubator at 37.5°C.

Procedure

In Vivo

1. Take a chick egg incubated for 4 to 8 days (Hamburger and Hamilton stage 24 to 35) and keep it with its blunt end up for 15 m in the incubator, to let the embryo come on top.
2. With the help of a blunt forceps make a hole (equal to a 25p piece) on the blunt end of the egg, removing the shell and shell membranes.
3. With a fine forceps expose the chorioallantois.
4. With the help of a Pasteur pipette put 2 to 3 drops of neutral red, stain on the chorioallantois near a blood vessel.
5. Incubate the embryo for 30-45 min.

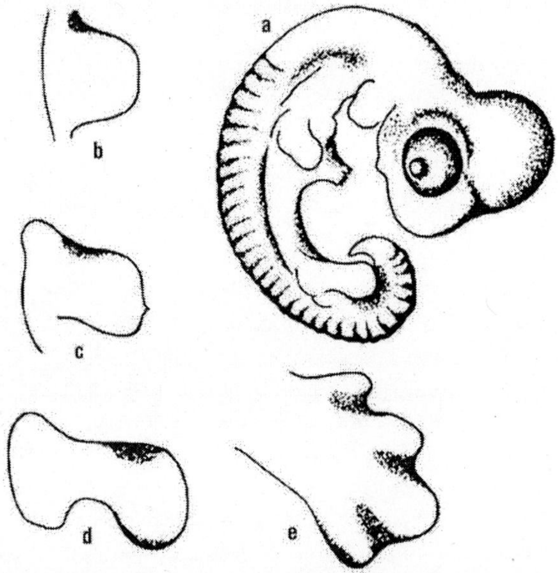

Fig. 10: Development of digits in the hind limb of a chick embryo. The area undergoing necrosis are shown by stepping.

6. Remove the embryo in a saline filled petri-dish and clean its all membranes, including amnion. Wash with saline twice.
7. Observe under the binoculars. Stained areas indicate cell death. For a better view of any organ, cut it away from the embryo and keep favorably in a watch glass with saline.

In Vitro for any Species

1. Fill a watch glass or cavity block with the neutral red solution and put into the incubator for warming.
2. Remove the embryo from the egg or uterus in pre-warmed saline in petri dish.
3. Remove all the embryonic membranes except the amnion.
4. Transfer the embryo into the watch glass and put into incubator for 30-45 minutes.
5. Transfer the embryo into saline, remove the amnion and wash twice.
6. Observe under the binoculars and sketch the diagram.

Observations

Cell death is seen in different organs at different developmental stages cell death in the limb can be seen from stage 24 to 35 in the chick.

> ***Note:*** *Other vital stains like methylene blue or Nile blue sulfate can be used instead of neutral red.*

CHAPTER 11

Transplantation of Regenerating Limb/Tail Blastema in *Anuran* Tadpole

Transplantation technique is very useful in developmental biology to find out whether the presumptive areas of the blastula are actually predestined for a specific part in the future development or to know the differentiating capacities of an organ and its functioning method. In present experiment technique is applied for regenerating blastema in toad tadpoles to study its autonomic self-differentiating capacity independent of stump influence.

The transplanted tissue grows in the new site. The embryo or larva from which the tissue is taken called the donor and the embryo/larva which receives the tissue is called the host. The transplanted tissue is called the graft. Transplantation can be done from one part of an embryo to another part (autoplastic), or to another embryo of the same species (homoplastic), or to another embryo of a different species (heteroplastic), or to another embryo of different genera (xenoplastic).

A transplanted tissue may grow to form a similar structure as its neighbouring region of the host or it may form a different structure different from its neighbouring region, depending on at what stage of development, the tissue is removed. If the tissue is taken from a blastula but not from early gastrula then it forms the same structure as its surrounding tissues in the host. After the gastrula stage, the tissues lose their ability to change. If a small part of the neural plate region of a late gastrula is transplanted to the epidermal region of the same embryo or another embryo it will form the neural cells. Thus after gastrulation, parts of the embryo get determined to form definite structures. It does not matter where they are transplanted. This is called determination. Same thing happens with regenerating blastema of toad tadpoles. Wherever the blastema is transplanted it will give rise only the lost part of the limb/tail.

Materials

For this experiment tadpoles of the toad **Bufo melanostictus** employed were at stage 30/31 according to Khan (1966) at the time of amputation, stereoscopic binocular, microscope, irridectomy scissors or a sharp blade. Sterile Ringer solution (0.7% NaCl) cotton, Petridishes instruments for histological study, MS 222 (1:4000) solution sharp and pointed needles and finger bowls.

Procedure

1. Select out 20-30 tadpoles of stage 30/31 and rear them in pond/tap water.

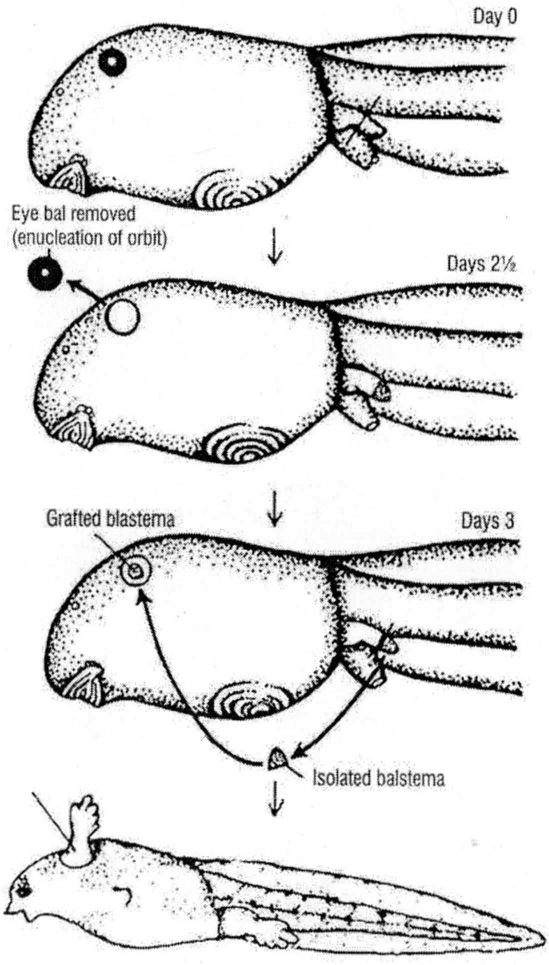

Fig. 11.1: Schematic drawing to show the procedure of auto-transplantation of shank level blastema into enucleated eye orbit in the tadpoles of *Bufo melanostictus*

2. Anaesthetize each tadpole in MS222 solution before amputation through shank.
3. On day 3 after amputation blastema is formed on amputated limb (stump).
4. After 2 and a half days the left eye orbit is enucleated by removing the eye ball in each tadpole. For this purpose a longitudinal incision is made in the skin of the dorsal side of the head above the eye; the eye ball is made free in the orbit by cutting the eye muscles with the help of a fine needle and transecting the optic nerve with a pair of a fine scissors. Now the eye ball including the lens thus separated is removed from the orbit by a pair of fine forceps through the dorsal incision.

5. The wound thus made is allowed to heal overnight during which time a fibrous vascular pad developed in the enucleated orbit.
6. On day 3 following limb amputation the regenerating limb is amputed by a fine blade under the stereoscopic binocular so as to isolate the blastema only. Care is taken to make me re-amputation level pass through the basal region of the blastema without including any part of the stump.

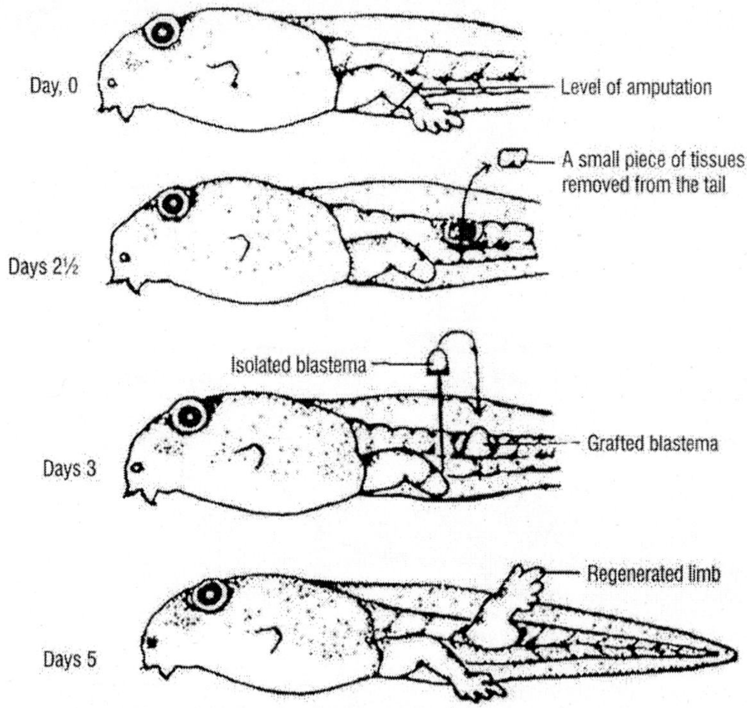

Fig. 11.2: Schematic drawing to show the procedure of autotransplantation of shank-level blastema into the incision made in the tail of tadpole.

7. The isolated blastema is then grafted autoplastically in the enucleated eye orbit of the same individual.
8. Similarly the isolated blastema can also be transplanted autoplastically or heteroplastically into the tail of the host of early stage tadpole. For grafting, a small incision is made in the eye orbit/tail and then blastema is gently inserted through this incision.
9. While crafting is being done care is taken to orient the blastema in such a way that its proximal basal end is towards the body axis and the conical distal end is directed away from this axis.
10. The operational procedures are shown in Fig. 11.1, Fig. 11.2 and 11.3.
11. Before these operations the tadpoles are narcotised in 1:L4000 MS222

solution.

12. During the operations the tadpoles are kept on a pad of filter paper moistened with MS222 solution.
13. After re-amputation and grafting each tadpole is kept immobile for 2-3 hours by keeping i partly immersed in diluted MS222 solution in a petri-dish. This allowed time for the graft to be taken. After this procedure tadpoles are transferred to tap water.

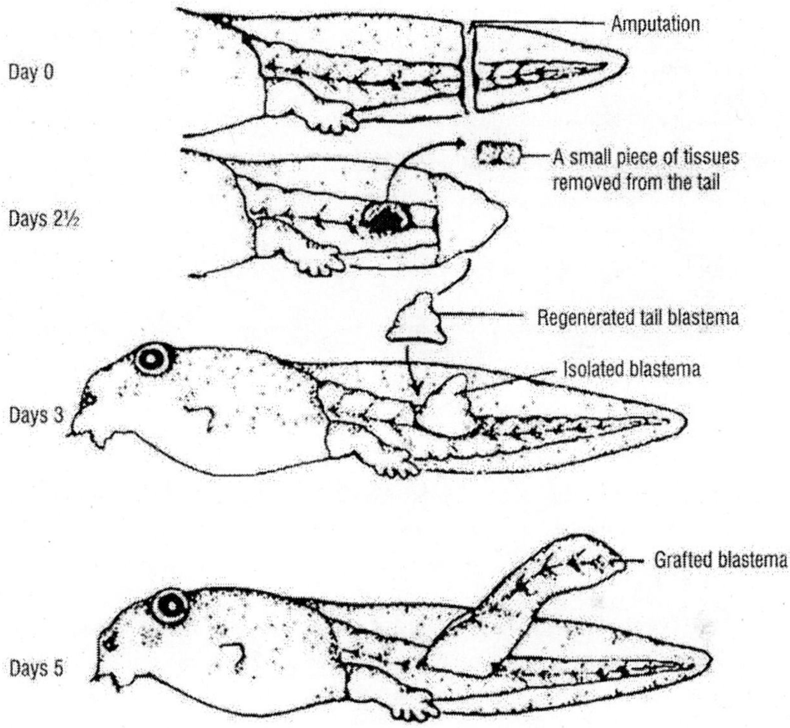

Fig. 11.3: Schematic drawing to show the procedure of autotransplantation of tail into the incision made in the tail of tadpole.

Observations

The blastemas grafted into the orbit/tail differentiated in this new location autonomously to give rise the limb regenerates (Fig. 11.1, 11.2 and 11.3). Histologically, the regenerates consisted of differentiated material and phalangeal cartilages, a few muscle fibres, connective tissues and epidermis.

CHAPTER 12

Experiments on Embryonic Induction

Embryonic induction is an interaction between one tissue (inductor) and another (responding tissue), as a result of which the responding tissue undergoes a change in its direction of differentiation. This is probably the single most important mechanism in vertebrate development leading to difference is between cells and to the organization of cells into tissues and organs.

The inductor may exert an **instructive** or a **permissive** influence on the responding tissue. The instructive influence leads to a restriction of the prospective potency of the responding cells and is therefore involved in a determinative process that limit the choices available to cells of the responding tissue to differentiate in various ways. The permissive type of inductive influence provides some kind of stimulus to the responding cells to differentiate in a direction for which they are already committed or determined but require certain conditions in which to do so, and these are fulfilled by the presence of the inductor.

In the chordates differentiation of many tissues during embryonic development is dependent on one of the two types of inductive influences. The first and perhaps the most important induction occurs during gastrulation when the chordamesoderm invaginates and comes to lie under the dorsal ectoderm. The chordamesoderm induces the overlying ectoderm to differentiate into neural tissue and ultimately form the entire central nervous system. The inductive action of the chordamesoderm on the ectoderm is of the instructive type so that the ectoderm can now differentiate only into neural tissue and give rise to various structures of the nervous system. Also, during gastrulation and thereafter the basic axial organization of the body is laid as a result of the interaction between the notochord (formed from the chordamesoderm) and the surrounding ectodermal, mesodermal and endodermal cells. Without the inductive action of the chordamesoderm neural tissue does not differentiate and the nervous system is not formed. This inductor when combined with prospective epidermal ectoderm of the early gastrula induces the latter also to differentiate into neural tissue. This phenomenon was discovered by Spemann and Mangold in 1924 in urodele amphibians who called it Primary Embryonic Induction. It has since been found to be true for all chordates.

The cells of the inductor (chordamesoderm) need not be in a living state to influence the ectoderm. Even dead cells also induce neutralization in the ectoderm. Actual physical contact between the inductor tissue and the responding ectodermal cells is not involved in induction. Inducing agent is some kind of a diffusible chemical which is probably released from the inductor cells and influences the overlying ectodermal cells.

Apart from the normal, natural chordamesoderm a number of other agents also induce neutralization in the gastrula ectoderm. They include embryonic, larval and adult tissues of a variety of animals from different phyla, many different chemicals, inert materials etc. Even sub-lethal injury to ectodermal cells may induce them to differentiate into neural cells. Whatever be the inducing agent it appears to be of a very non-specific nature but the response of ectodermal cells is always very specific.

The inductive influence of the chordamesoderm is regionally specific. The anterior part of archenteric roof induces the formation of head and forebrain structures and is called the archencephalic inductor. The middle part of the archenteric roof induces the differentiation of midbrain structures and is known as the deuterencephalic inductor. The posterior part of the archenteric roof causes the formation of spinal cord and caudal structures and is designated as the spino-caudal inductor.

Some animal tissues or their extracts produce regionally specific inductive effects comparable to those of the various antero-posterior parts of the normal inductor. Some of these heterogenous inductors are archencephalic, others are deuterencephalic or spino-caudal. Some are purely mesodermal inducing ectodermal cells in differentiate into mesodermal structures. Chemical analysis of some of these heterogenous inducing agents has shown that the inducing substance in the purely archencephalic inductors (i.g. guinea pig kidney) is of the nature of ribo-nucleoprotein and that of the purely mesodermal inductors (i.e. guinea pig bone marrow) is some kind of protein. In both cases the integrity of the protein is essential for producing inductive effect. It is still not known whether similar one, both or more factors are produced by the different regions of the natural inductor (chordamesoderm) during embryogenesis.

The ectoderm must be in a physiological state competent to respond to the inductive influence of the chordamesoderm. This competence arises gradually as blastulation is completed and disappears as the process of gastrulation comes to an end. The appearance and disappearance of competence of the ectoderm to respond to induction from the chordamesoderm is an autonomous property of the ectoderm and is a function of time.

The neural inducing property is present in the cortex of the fertilized egg. This is localized in the grey crescent region and increases in potency during the first few divisions after fertilization. Irradiation of the grey crescent region with U-V rays after fertilization destroys this property. Injection of material from the germinal vesicle of a mature oocyte into such irradiated egg restores this property. It means that the inducing substances, whatever be its nature, is already present in the cytoplasm of the egg and it becomes localized in the cells of the prospective chordamesoderm which form in the grey crescent region.

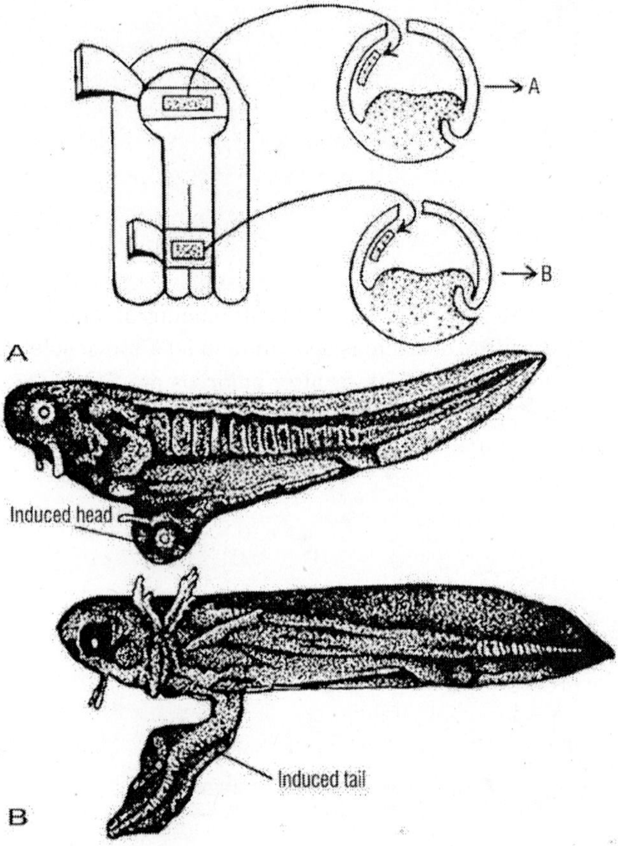

Fig. 12.1: The separation of the organizer into head and trunk organizer. The upper figure illustrates a neural plate stage in which the anterior neural plate (A) has been cut and the underlying roof of the archenteron has been exposed.-This region of the archenteron roof is then cut out and transferred to the blastocoel of an early gastrula. The embryo (A) which develops from the gastrula possess an induced head. In the upper figure posterior neural plate (B) was cut out and the underlying roof of archenteron transplanted to the blastocoel of an early gastrula, an induced secondary trunk projects from the side of the embryo (B).

The responding ectodermal cells synthesize new mRNAs after the induction has occurred. Without these new RNAs they do not differentiate into neural tissue.

Based on the specific inducing properties of the heterogenous tissues it has been suggested that gradients of two inducing agents may be involved in the entire process of primary embryonic induction and formation of the regionally specific parts of the nervous system and other axial structures. One of these agents may be neutralizing factor with its highest concentration along the mid-dorsal line from anterior to posterior end and decreasing ventrally on either side.

The other agent may be a mesodermalizing factor with its highest concentration at the posterior end of the archenteric roof decreasing both anteriorly and latero-ventrally. The relative proportions of the two factors in different regions along with antero-posterior axis and latero-ventrally induce the formation of various regionally specific structures in the embryo. Experiments with extracts of archencephalic and mesodermal heterogenous inductor tissues have provided support to this model of Primary Embryonic Induction in amphibians.

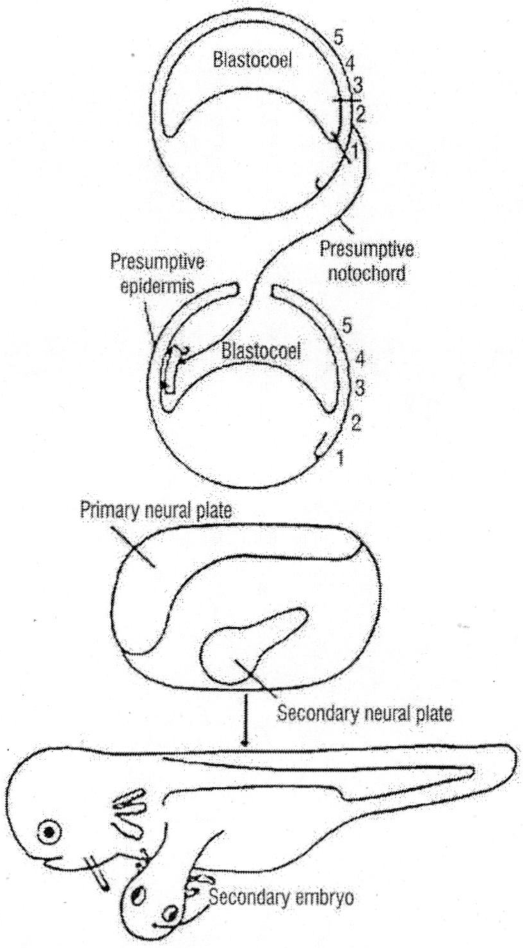

Fig. 12.2: The formation of a secondary head after transplantation of regions 1 and 2 in the early gastrula. Top region 1 and 2 are transplanted into another gastrula in such a way as to bring them into contact with the presumptive epidermis. The second gastrula forms a primary neural plate and a secondary neural plate. The Final result (bottom) is an embryo with two head. It is important to note that the secondary embryo contains only a very small amount of tissue contributed by the transplant. Most of the tissue comes from the host.

Materials

Early embryos of required stage 50% NAM (normal amphibian medium Slack & Forman, 1980) Embryos fixed in 4% para formaldehyde (PF) or Romeis Fixative (25 ml saturated Hg Cl, 20 ml 5% TCA, 15 ml 37% formaldehyde).

Stereo-binocular microscope. Watch-glasses or Petri-dishes. Drawing note book etc.

Procedure

1. Examine the embryos by the aid of the binocular microscope under an appropriate magnification. Sort them into definite stage numbers in separate glass dishes & label them.
2. Transplant the anterior part of archenteric roof to early gastrula (host A) and
3. Also transplant the posterior part of the archenteric roof to the B host of early gastrula stage.
4. Allow them for further development in the NAM (normal amphibian medium).

Observations

The inductive influence of the chordamesoderm is found to be regionally specific and the regional specificity is imposed on the induced organs by the inductor. The anterior part of the archenteric roof induces head organs and may be called a head inductor the posterior part of the archenteric roof induces trunk organs and tail buds and may be called the trunk or spino caudal inductor. The head inductor further differentiated into the archencephalic inductor, inducing forebrain, eyes and nose rudiments, and the deuterencephalic inductor, inducing hindbrain and ear vesicles.

CHAPTER 13

Teratological Experiments on Chick and Frog Embryos

Teratology is the study of toxic responses in offspring to environmental agents or drugs following exposure of mother at any time from conception to birth. Teratology includes any birth defect, morphological, biochemical or behavioural, induced by an embryo-toxic agent at any stage of gestation. It is primarily concerned with factors that interfere with the developmental differentiation.

Teratogenic agents are able to initiate abnormal development by number of mechanisms such as gene mutation or chromosomal abnormality, altering nucleic acid function, membrane characteristic, energy source and inhibiting the enzyme. The teratogenic agent initiates one or more of above mechanisms resulting in abnormal embryogenesis. This in turn leads to pathways that seems to effect localized morphogenesis of functional maturation or cause other imbalances in growth and differentiation, which ultimately leads to abnormal developments in embryo. The study of teratogens has significantly contributed our understanding of normal development.

Materials

Fertilized eggs of Hen/ Frog, incubators, petridishes, saline solution, different teratogens like thalidomide, parathion, cadmium salts, retinoic acid, sodium fluoride etc. Bouin's solution alcohol xylene etc. Forceps, scissors, needles, dropper and microscope.

Procedure

1. Fresh embryonated eggs are wiped clean and incubate at 38°C for 17 hr.
2. A hole is made with the help of a cutting needle and then a window of small diameter is cut open with the help of iris scissors.
3. The blastoderm is visualized as a disc floating uppermost on the yolk covered by a thick layer of albumin.
4. The injections are made into the sub-blastodermal space (that ensures drug delivery into the embryo) at 17 hr. using a 26G needle and a tuberculine syringe in a constant volume of 0.06 ml.
5. To avoid injury and desiccation of the blastoderm the thin albumin is carefully pipetted out, from the side of the yolk without disturbing the embryo and the window is sealed with cello tape.
6. The eggs are further incubated for 40 hr. At the end of this period the window is enlarged till the level of the albumin. The vitelline membrane along with

blastoderm is carefully taken out and shifted to petridish filled with water.

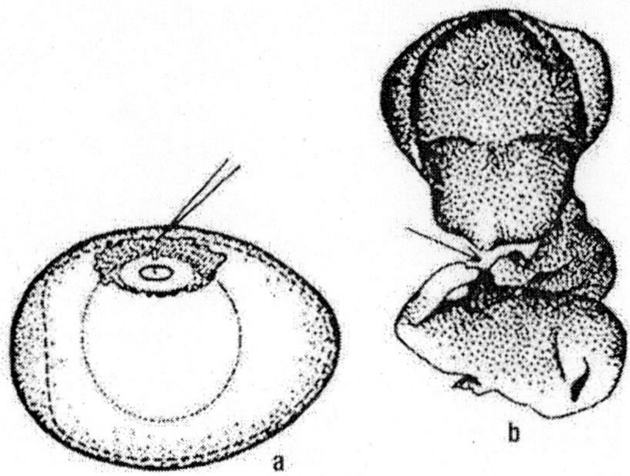

Fig. 13.1: Production, of anterior symmelia (sympteria) by action of a teratogen. (a) Teratogenic solution is dropped on a 15 somite-stage chick embryo (42 hrs.). (b) 8-days-old chick embryo treated by cadmium showing fusion of the wing buds (arrow). The neural tube is not interrupted at this level.

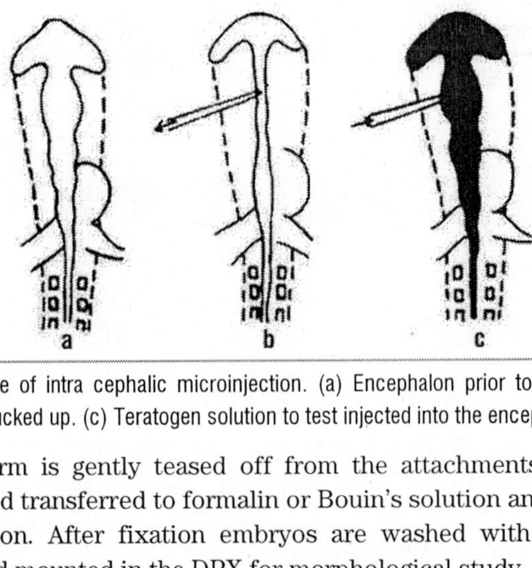

Fig. 13.2: Technique of intra cephalic microinjection. (a) Encephalon prior to operation. (b) The encephalic fluid is sucked up. (c) Teratogen solution to test injected into the encephalon.

7. The blastoderm is gently teased off from the attachments to the vitelline membrane and transferred to formalin or Bouin's solution and kept for 24 hrs. in this solution. After fixation embryos are washed with water and then processed and mounted in the DPX for morphological study.
8. The simplest method consists in dropping a teratogenic solution on young chick embryos. Eggs are prepared according to Fig. 13.1 and Fig. 13.2. A teratogenic solution is then deposited directly on the embryos, with or without removal of vitelline membrane.

One can produce various preparations at one time and thus select precise developmental stages for administering the teratogen.

Results

1. 1 hrs. stage i.e., primitive streak stage is period of great proliferation and differentiation and therefore maximum teratogenicity occurred.
2. Injection of thalidomide produced mal-development.
3. Parathion in chick embryos of 1, 2, or 4½ days of incubation with doses equivalent to those utilized in agriculture. The treated embryos showed growth retardation associated with limb malformations. Substances injected into the subgerminal cavity induce high percentages of malformations.
4. Cadmium salts are highly teratogenic when administered by the direct routes to 2-day old chick embryos.
5. The albumen has a greater ability to accumulate cadmium than yolk.
6. When a single dose (0.7, 0.35, 0.175 gm/embryo) of retinoic acid dissolved in DMSO is injected to chick embryo of 3 days incubation through a window in the shell. Growth of treated embryos is severely inhibited particularly morphogenesis and differentiation of* the head and limbs are adversely affected.
7. On the administration of sodium fluoride at the concentration of 7g in one ml of autoclaved chick ringer. Teratogenic effects like shortening t.)f beak, reduced cornification of the tips of toes, naked skin due to the loss of feather germs and blackening of coverts and feather germs etc. resulted.

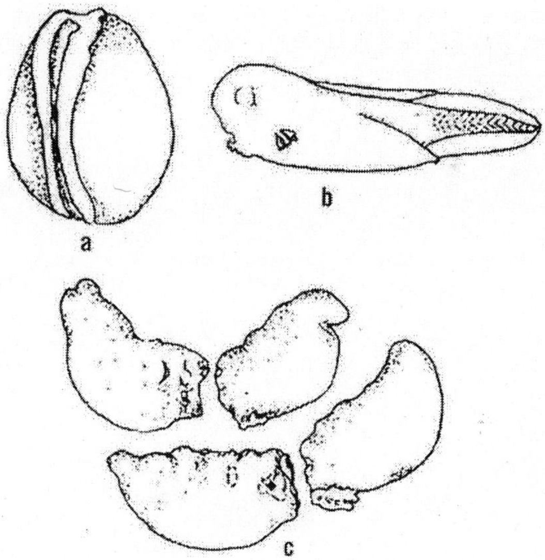

Fig. 13.3: Teratological experiments on frog embryos. (a) Frog embryo, Neurula stage. (b) Normal development at room temperature 30°C. (d) Frog's abnormal embryos after keeping neurula at 5°C for 12 hrs.

8. Physical changes such as temperature can produce mal-development.

Temperature variations (the temperature of the injected saline) from the normal range of 37°C to 39°C produces defect in development of neural tube. Rise in temperature (50°C) produces mainly abnormal anterior end and retarded posterior end growth while decrease in temperature (10°-20°C) leads to anterior end abnormality and posterior end nonfunctional. Hypothermia produces predycardia in chick embryo which may be due to changes in cell membrane potential as Na^+, K^+, ATPase activity is directly related to variation in temperature. Whereas hypertherma causes endothelial damage and endothelial gaps leading to vascular leakage. This vascular leakage leads to interstitial oedema which may interfere with the neural tube development.

Similar abnormal developments have been found in Frog's embryos when kept at 5°to 10°C for 12 hrs. (Neural stage. Fig. 113)

CHAPTER 14

Study of Tail Regeneration

Regeneration of Tail in Tadpoles

The replacement of lost parts of an organism is called regeneration. Regeneration is predominant in amphibians among vertebrates. An amphibian tadpole can regenerate tail, limbs, external gills, jaws and even part of eyes.

Amputation of tail is followed by wound healing, dedifferentiation, cell proliferation, cell movements, re-differentiation, morphogenesis and growth of regenerates. All these processes also occur during ontogenetic development in the embryonic and larval stages. For study of these events the amphibian tadpoles of early stage are most favourite animals.

Materials

Binocular microscope/dissecting microscope, Frog or toad tadpoles of 4-5 days after hatching. Scissors, sharp knife or razor/blade finger bowls, droppers, 1:4000 MS222 solution petri dishes etc.

Procedure

1. Select the tadpoles of desired developmental stages with prominent tail fin. Anaesthetize the tadpole by placing in 1:4000 solution of MS 222 for one or two minute.
2. Transfer the immobilized tadpoles on a bed of wet filter paper.
3. Under the dissection microscope, make a sharp and neat cut using the razor or blade, avoid crushing the tail tissues.
4. After tail amputation transfer the tadpoles to the tap water in finger bowls. The tadpoles revive within few minutes and resume swimming.
5. After wound healing (12 hrs) fed the tadpoles with half boiled spinach. 1

Observations

1. Observe the process of wound healing, and the process of regeneration under the binocular microscope.
2. Measure the tail length of each tadpole every alternate day, calculate mean of tail length of each group of amputated tadpoles.
3. Record the observations until the whole tail is regenerated.
4. Fix few amputated tadpoles on day 1, 2, 3, 4 & 5 in Bouin's solution for morphological and histological observations.
5. Plot a curve between the length of regenerate and day after amputation.

   Tail regeneration in frog/toad tadpole.

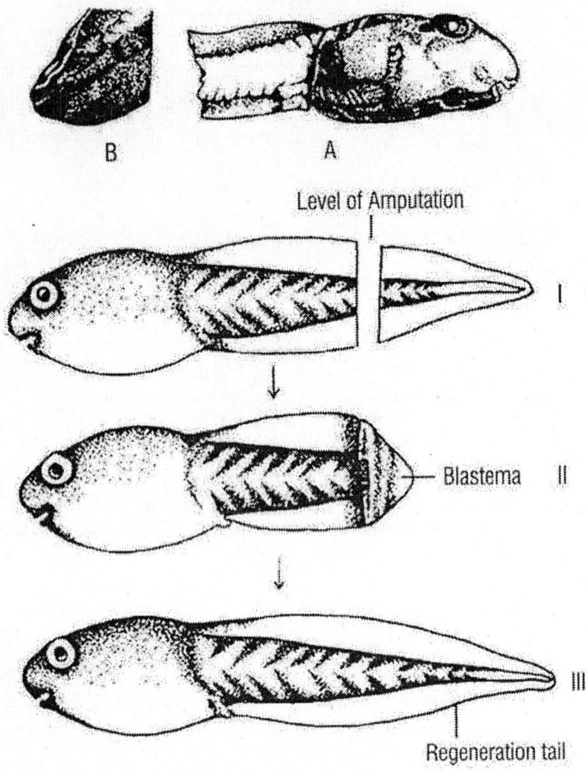

Fig. 14: Tail regeneration tadpole of *Bufo melanostictus*. A and (i) Showing level of amputation. (ii) Blastema is formed on the tail stump. B and (in) Showing regenerated tail.

Tail regeneration in frog/toad tadpole

Date	Day after Amputation	Level of Amputation	Morphological Observation	
			Shape	Length

CHAPTER 15

Regeneration of the Amphibian Limb

Regeneration is defined as restoration of the lost part of an organ or rèconstitution o the whole body from a fragment. It is a developmental phenomenon occuring during post-embryonic period in an already formed and functional' organism. Regeneration involves all those fundamental processes including cell proliferation, cell movements, morphogenesis, histogenesis and growth which occur during ontogenetic development in the embryonic and larval stages.

Among vertebrates the amphibians have been the most favourite animals for study of generation as they exhibit this capacity to a remarkable degree.

In the young anuran tadpoles ability to regenerate hind limbs amputated at any level is high but gradually declines proximo-distally along the limb axis ultimately disappears throughout the limb before metamorphosis.

Injury or amputation is must for regeneration. The immediate effect of amputation is of course, the exposure of interior cells of the limb, many of which are killed by the operative trauma bleeding occurs from the damaged blood vessels and a clot forms over the cut surface. Epidermal cells from the stump migrate distally and cover the wound surface (wound healing). The epithelium is at first only a single layer of cells, but the proximal epidermis proliferates id shifts toward the wound area so that an apical ectodermal cap several cells in thickness accumulates. It is almost equivalent to the apical ectodermal ridge (AER) which forms on the) of the limb bud during ontogeny. Meanwhile, subjacent to the cap, the tissues undergo a dramatic change that is referred to as dedifferentiation. The intercellular matrix material of cartilage and other connective tissues are digested away, releasing the cells from their tissue organization. Sheath cells of Schwann dissociate from their nerve fibers, and multinucleate fibers of skeletal muscle break into fragments. Nucleated fragments of muscle lose their myofibrils and join the other released cells in forming a mass of proliferating tissue that accumulates beneath, the AER as the blastema. Cells of the blastema are histologically, cytologically and immunologically similar. This process is known As "dedifferentiation" through which blastemal cells originated from stump tissues.

After the blastema has attained a certain size morphogenesis of the regenerate begins. The conical blastema flattens distally to form a foot paddle, rudiments of toes appear, various limb segments are demarcated and the regenerate assumes the shape of a miniature limb. Internally, the re-histogenesis of tissues is also initiated. The next phase is of rapid growth of miniature regenerate which ultimately becomes similar to the normal intact limb.

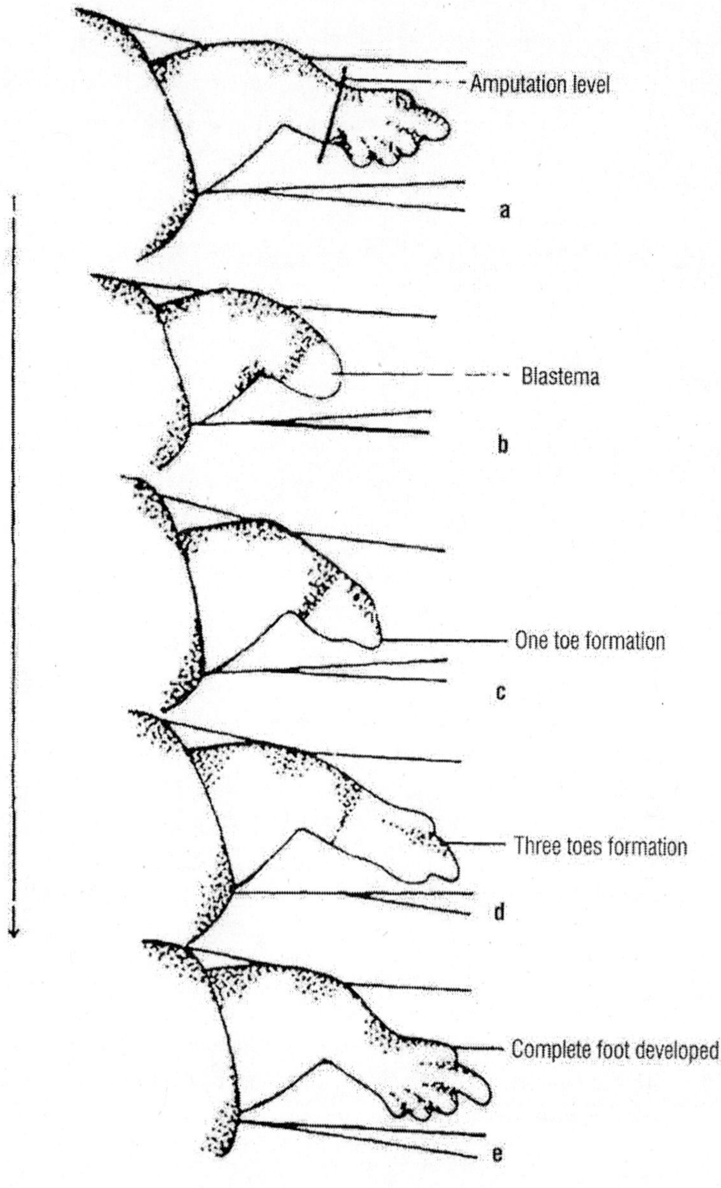

Fig. 15: Limb regeneration in tadpole of Bufo melanostictus. (a) Amputation level. (b) Blastema. (c) One toe formation. (d) Three toes formation. (e) Complete foot develops.

Materials

Frog or toad tadpoles of 3-5 days after hatching, scissors, sharp knife or blade, Finger bowls. Droppers = 1:4000 MS222 solution. Petri dishes and Binoculars etc.

Procedure

1. Select the tadpoles of desired developmental stages (of early stages). Anaesthetize the tadpole by placing in 1:4000 MS 222 solution for one or two minutes.
2. Transfer the immobilized tadpoles on a bed of wet filter paper.
3. Under the dissection microscope, make a sharp and neat cut using the razor or blade through shank region. Avoiding crushing the limb tissues.
4. After wound healing (about 8-12 hrs) feed the tadpoles with half boiled spinach.

Observations

1. Observe the process of wound healing and the process of regeneration under the binocular microscope.
2. Observe the morphological changes in regenerating blastema/limb. Count the number of toes in the regenerating foot and sketch the diagrams of regenerate alternate days.
3. Record the observations until the whole limb is regenerated.
4. Fix few amputated tadpoles on day 1, 2, 3, 4, 5, 6, 7 and 15 in Bouin's solution for morphological and histological observations.

Table 15.1. Limb regeneration in frog/toad tadpoles.

Stage at amputation	Level of amputation	No. of limbs amputed	Limbs generating No. (%)	Morphological quality of Regeneration Regenerates with					Remark
				5 Toes	4 Toes	3 Toes	2 Toes	1 Toe	

CHAPTER 16

Lens Regeneration (Wolffian Regeneration)

Lens regeneration is a special kind of regeneration occurring in amphibian tadpoles. It was discovered by WOLF in 1935. The regeneration of a new lens is called Wolffian regeneration. Amongst other vertebrates lens regeneration has been observed in the chick embryos and even in newly born mice babies by Jangir et al (1995).

Materials

Frog or Toad tadpoles of 3-5 days after hatching, scissors, forceps, sharp knife or sharply pointed needles, finger bowls, droppers, 1:4000 MS222 solution, petri dishes, binocular microscope, filter paper and cotton.

Procedure

1. Select the tadpoles of desired developmental stages (of early stages). Anaesthetize the tadpole by placing in 1:4000 MS222 solution for one or two minutes.
2. Transfer the immobilized tadpoles on to a bed of wet cotton or filter paper.
3. Under the dissection microscope or binocular make a small and fine longitudinal slit in the cornea. The lens capsule is seized by means of fine forceps inserted in the corneal opening by fine needle.
4. Intact lens is easily extracted through the incision.
5. Operated tadpoles are then reared in to finger bowls containing water.
6. After wound healing (about 8-12 hrs) feed the tadpoles with half boiled spinach.

Observations

1. Observe the process of wound healing and the process of regeneration under binocular microscope.
2. Sketch the diagrams of regenerate on alternate days.
3. Preserve 4-5 tadpoles in Bouin's solution for histological study every day.
4. Cut the serial sections of each day preserved groups of tadpoles and observe.

Note: *The original lens is developed from epidermal ectoderm. But the regenerating lens is formed from iris (neuroectoderm). Thus, in Wolffian regeneration, the lens regenerates from a tissue different from that gave rise to it during embryonic development.*

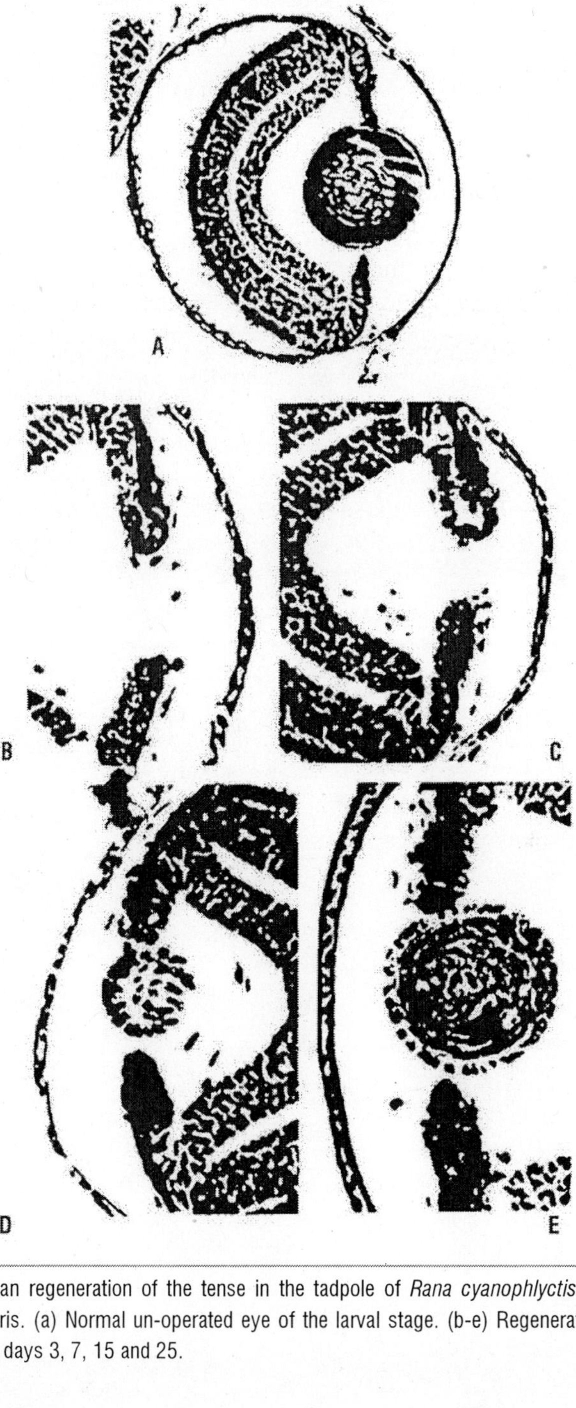

Fig. 16: Wolffian regeneration of the tense in the tadpole of *Rana cyanophlyctis* from the dorsal margin of the iris. (a) Normal un-operated eye of the larval stage. (b-e) Regeneration of lens seen respectively on days 3, 7, 15 and 25.

CHAPTER 17

Metamorphosis in Amphibians

The term metamorphosis is derived from a Greek word metamorphoun which means transformation. In embryology it is defined as transformation of the larva into adult during which many structural as well as physiological changes take place.

It is known that thyroid glands play key role in amphibian metamorphosis. Thyroxine, (tetraiodo tyrosine) a principal product of the thyroid gland is a derivative of the phenolic amino acid Thyronine. This hormone is responsible for the acceleration of the metamorphic events in the frog tadpoles. Thyroid gland collects iodine from the blood stream and builds it into the molecules of thyroid hormones. The production of thyroxine thus depends on the iodine supply available to the animal. Various anti-thyroid drugs are known which suppress the thyroid activity e.g. thiourea, potassium perchlorate etc. The present experiment demonstrates role of thyroid gland in amphibian metamorphosis.

Materials

Fully developed tadpole larva and newly metamorphosed froglets, a low-power binocular microscope with spot-light lamp, Ocular micrometer, Pair of blunt and pointed forceps, spatulae, Watch-glasses and small glass dishes. Pads of gauze and cotton. Drawing note-book, drawing set. Ruler or graph paper for measuring gross lengths, MS 222 or chloretone (concentrations of 1:4000 or 1:2000 respectively in tap water). Ethyl alcohol or Bouin's solution.

Procedure

1. Prepare various concentrations of thyroxine hormones preferably 10-7 or 10-8 mg/ml.
2. Prepare weak solution (of various concentrations) of potassium perchlorate.
3. Collect frog or toad tadpoles of pre metamorphic stages using a nylon net.
4. Narcotize the tadpoles in 1:4000 solution of MS 222 (Sandoz) and identify the stage of development.
5. Select 5 groups each of 20 tadpoles.
6. Transfer tadpoles to the bowl filled with two different concentrations of thyroxine and another group to the different concentrations of potassium perchlorate.
7. Rear the control group in tapwater.

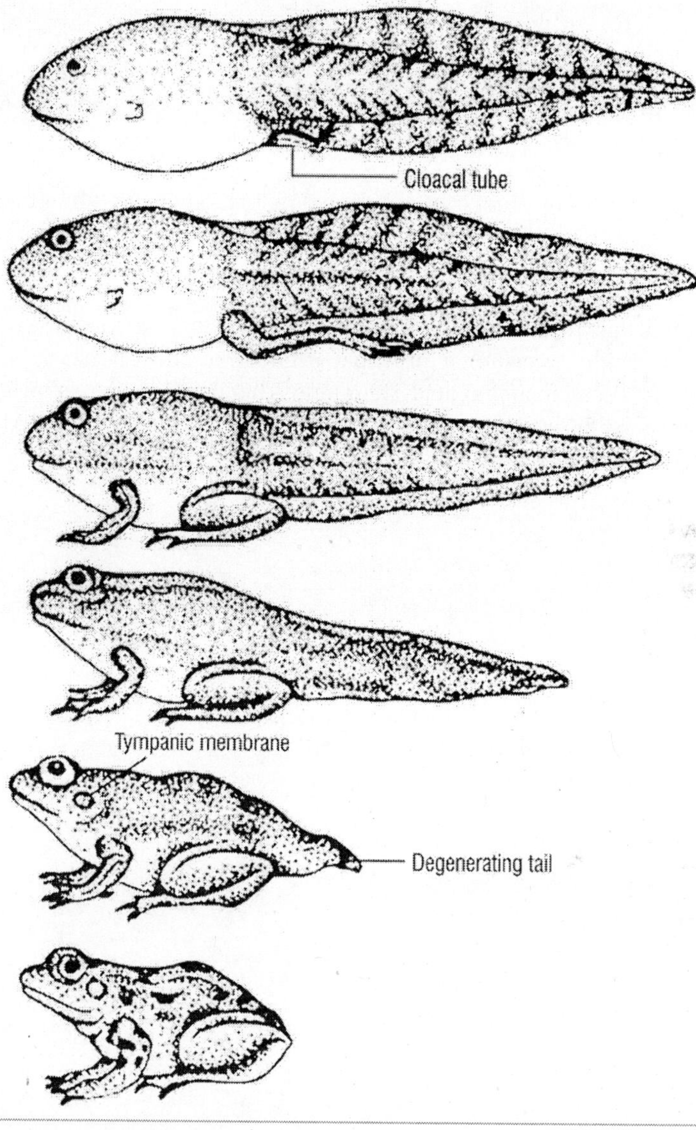

Fig. 17.1: Late larval and metamorphosing stages of *Rana cyanophlyctis*.

Observations

1. Thyroxin accelerates metamorphic changes whereas the tadpoles reared in potassium per chlorate solution show delayed metamorphic changes.
2. Fix the metamorphosed tadpoles in Bouin's solution and note the time of metamorphosis.

Note: Metamorphosis in anuran amphibians is a process which causes drastic changes in the structure of the body as shown in several events encountered in the external features as well as in the internal features recorded in the different organs and systems. By this process the completely aquatic larval tadpoles are transformed into semi-aquatic terrestrial froglets. They become adapted to stop respiration by gills and be able to breathe atmospheric air. The kind of diet also changes. Shrinkage in the overall length of the head trunk during the metamorphic is well observed. The metamorphic stages are not adapted to feed since the widening of the mouth opening is accompanied by re-modelling of the skeleton of the jaws.

Thyroxine hormone is responsible for the acceleration of the metamorphic changes in the frog tadpoles and anti-thyroid drugs like potassium perchlorate delays the metamorphosis.

Changes. Shrinkage in over all, length of the head and trunk.

1. Anal canal disappears.
2. Horny beak is lost.
3. Wide froglike mouth is formed.
4. Each forelimb emerges.
5. Hind limbs grow to 1 cm long 2 cm; 4 cm; etc.
6. Tail is reduced to half its initial length.
7. Eardrums appear.
8. Sides of body are pinched. in.

Review the notes and summarize your findings. Make a table in which record number of days and controls show above changes.

Precautions

- **Step 1.** Experimental and control groups should be examined, fed, and cleaned every 2nd or 3rd day.
- **Step 2.** From time to time you may find that some of the tadpoles have died. If so, remove them, but do not replace them. (Why not?) Should the death rate seem to be too high in an experimental group, it might be a good idea to reduce the concentration of the hormone being used, consult your teacher about this.
- **Step 3.** Examine tadpoles minutely. Make notes on what you see, and compare with notes from preceding days. When the hind limbs seem to be growing actively, measure them from time to time. Once the tail has begun to shorten, measure it also. Watch for the appearance of the *forelimbs and of the eardrum. Examine the mouth. Can you tell when the horny beak of the tadpole is lost? Follow the changes in the shape of the body. (It might help to make sketches in your notebook). Note when the tadpoles leave the water and rest on the stone you have provided for them.
- **Step 4.** Whenever you have a chance, compare your tadpoles with those being raised by other squads. Are others metamorphosing more or less rapidly

than yours? Are changes occurring in the same sequence in all groups?

Table 17.1: Summary of some metamorphic changes

System	Larva	Adult
Locomotory	Aquatic, tail fins	Terrestrial, tailless tetrapod
Nutritional	**Herbivorous:** Long spiral gut-intestinal symbionts, small mouth, horny jaws, labial teeth	**Carnivorous:** Short gut-proteases; large mouth-long tongue
Respiratory	Gills, skin, lungs; larval haemoglobins	Skin, lungs; Adult haemoglobins
Circulatory	Aortic arches; aorta; anterior, posterior and common cardinal veins	Carotid arch; systemic arch; angular veins
Nervous	Lack of nititating membrane, porphyropsin, lateral line system - Mauthner's neurons	Development of ocular muscles, nititating membrane, rhodopsin loss of lateral line system degeneration of Mauthner's neurons; tympanic membrane
Excretory	Mesonephros: largely ammonia some urea (ammonotelic)	Mesonephores: largely urea, high activity of enzymes of orithine-urea cycle (Ureotelic)

CHAPTER 18

Metamorphosis in Insects

Hormonal Control of Metamorphosis in Insects

In insects as in amphibians, the metamorphosis of larva to pupa and then adult is under the control of hormones. Larval and pupal molts in holometabolous* insects and nymphal molts of the hemimetabolous forms are all controlled by hormones. Periodically, neuro-secretory cells in the brain release a brain hormone called ecdysiotropin. This hormone stimulates (through corpora cardiaca) the prothoracic gland to secrete the moulting hormone (=ecdysone) and probably inhibits the secretion of juvenile hormone (=neotenin) by corpora allata. Ecdysone stimulates growth and causes the epidermis to secrete a new cuticle, initiating the moulting process. In most of the insects as in housefly**, the corpora cardiaca, prothoracic glands and corpora allata combine to form a single ring gland (Wiesman's ring) located in the fifth and sixth segment. So a third instar larva ligatured anterior to the 5th segment underdoes pupation posterior to the ligature only similarly to a ligature posterior to segment six allows pupation of anterior segments only. Ligatures on either side of ring gland prevent poupaption of both anterior and posterior portions. Pupation is completed in about six hours in housefly.

** To observe normal metamorphic changes in Butterfly. (Fig. 18.1)*

*** For experimental purpose take housefly larvae which are easily available in your 18.2)*

Materials

Housefly larvae i.e. maggots; Filter paper, Fine white cotton thread, Petri dishes, Dissection instruments, Binoculars.

Procedure

1. In the vicinity of cow/buffalo dung examine the housefly larvae and collect active third instar larva. They are of cream colour. Place in a petri dish on damp filter paper.
2. Make a loose knot of thread.
3. Allow the larva to crawl through the knot until it has reached the desired position. The anterior end of larva is narrow and with black mouth parts.
4. As soon as the larva reaches desired position quickly tighten the knot to constrict the larval body and secure it with another knot. Take care not to cut up the larva.

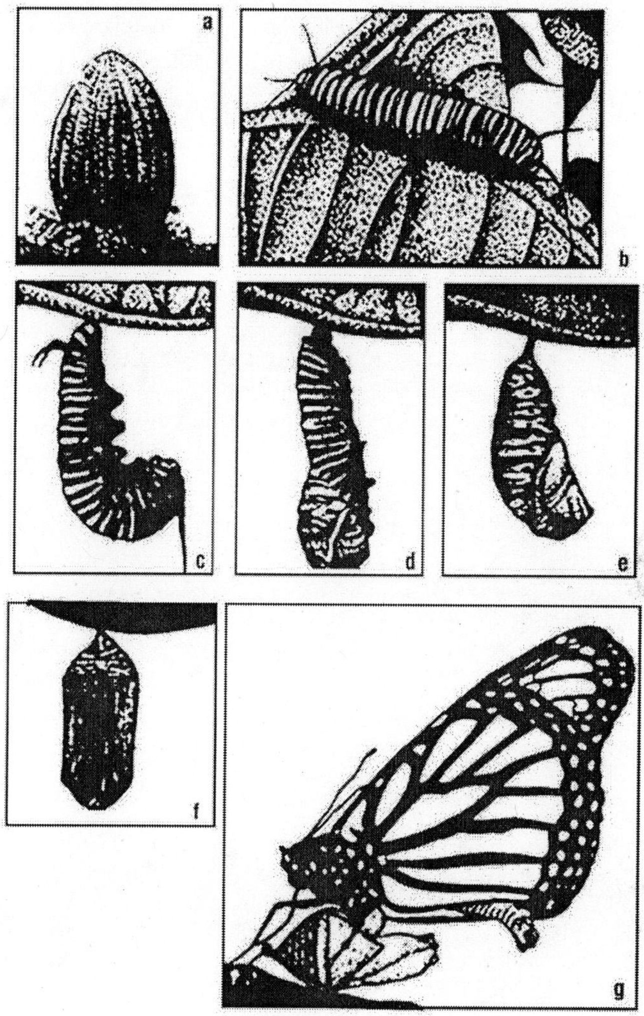

Fig. 18.1: Metamorphosis in large white butterfly (Pieris brassicae) (a) Egg. (b) Caterpillar. (c-e) Chrysalis. (f) Pupa attached to a twig. (g) Butterfly.

5. Leave the larva with knot in a covered Petri dish lined with damp filter paper, at room temperature.
6. The experimental larvae may be ligatured at different body segments.
7. Take a control group, without thread ligature to study metamorphosis in all the segments.
8. The experimental larvae should be ligatured as follows:
   a) **First group:** ligature anterior to segment 5;
   b) **second group:** ligature posterior to segment 6;

c) **third group:** ligature anterior to segment 5 and posterior to segment 6;

d) **fourth group:** no ligature, to serve as controls.

Prepare at least 10 larvae for each group.

a

b

c

d

e

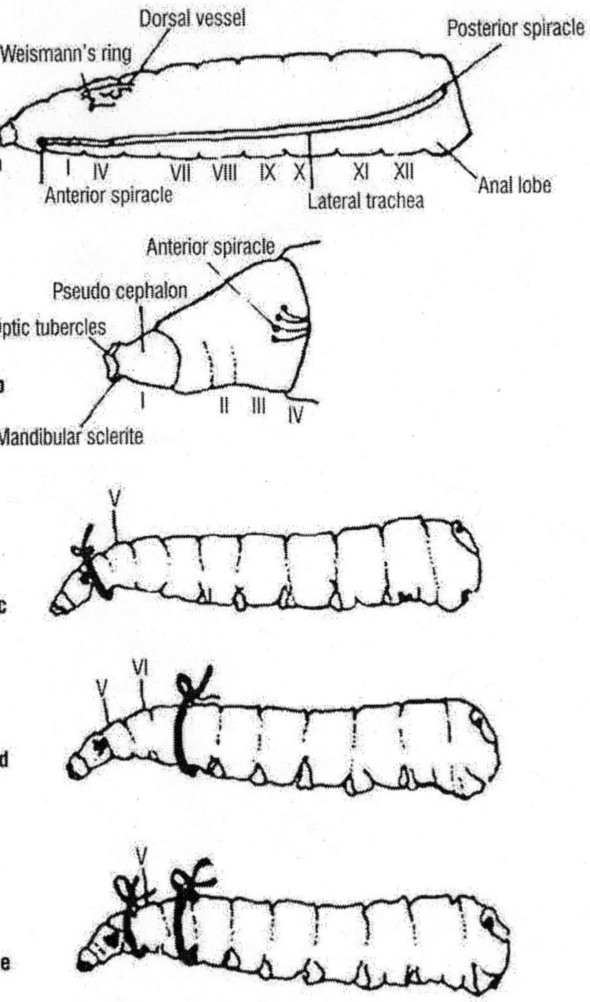

Fig. 18.2: Third instar larva or full grown maggot of housefly: Notice the position of Weismann's ring in V segment (a). Anterior end enlarged to show the mouth parts and opening of the spiracle in the third segment. c, d and e. Show the knoting to constrict the larva. Ligature anterior to segment 5 (c). Ligature posterior to segment 6 (d). Ligature anterior to segment 5 and posterior to segment 6. No ligature (a).

Results

1. Examine the larvae daily for pupation. Keep a record.

2. Note down the pupated body portion, that is, part which becomes rich dark brown. Make appropriate drawings.
3. Interpret the results with reference to the role of hormones in the metamorphosis.

Notes

1. Some larvae die due to the effects of ligature.
2. Many experimental larvae show complete pupation despite ligatures. This is particularly true for larvae ligatured 25 or less hours before pupation. This is due to moulting hormone having reached the whole body before ligaturing.
3. Some larvae do not pupate perhaps because the ring gland is injured.
4. Oxygen supply plays an important role in tanning of puparium.

CHAPTER 19

To Study the Early Development of a Fish

It is very desirable that the student of embryology become acquainted with the early stages of normal fish development. Such a study can be made on any fresh water fish available in the local ponds of water.

Materials

Adult fishes or spawn of the fish, Insect collection net, Fish tank, Glass troughs, Aquarium forceps, Droppers, Blotting paper, Formalin solution, 1:4000 MS 222 (sandoz), Bouin's fluid stereoscopic binocular microscope. Mixed dry food and worms, Watch glasses, petri dishes etc.

Procedure

Around 50 fishes may be kept in 15 gallon fish tank and may be fed on alternate days with worms and mixed dry food. If light, temperature, and food are properly regulated a good portion of the females in such a tank will produce eggs each morning shortly after dawn, and by 8:00 to 9:00 A.M. Many will be in early cleavage stages. The eggs are normally fertilized externally after a brief courtship by the male, just before or at dawn. The eggs (1 to 80) remain attached to the female for many hours unless they are eaten off by other fish. They may be picked off of the female with forceps and placed in glass troughs with some of the aquarium water, where they will develop normally through hatching (about 6 days). Following is a condensed schedule of early development of this fish egg. If material is available the student should study it.

Observations

1. Eggs becomes translucent after 10 minutes of fertilization.
2. First cleavage within 1 hour. Cleavage meridional; nucleoli visible.
3. Second cleavage at 1½ hours, at right angles to first. Meridional cleavage with equal blastomeres.
4. Third cleavage at 2 hours, parallel to first. Blastodisc rectangular.
5. Fourth cleavage at 3 hours, parallel to the second. Central blastomeres smaller.
6. Fifth cleavage at 3½ hours. Unlike most teleost eggs this does not give rise to layers of cells.
7. Sixth cleavage at 4 hours. Cleavages no longer synchronous;

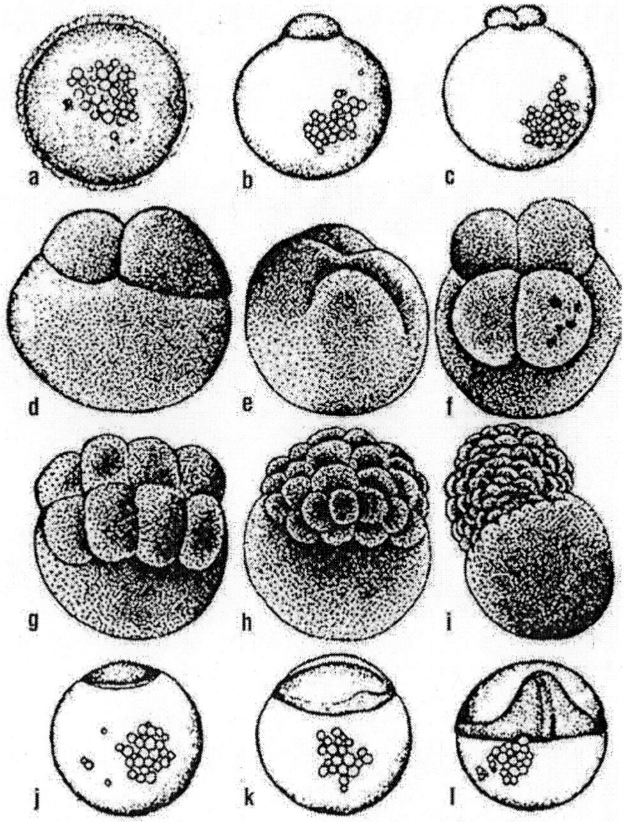

Fig. 19.1: Embryogenesis of a fish, Fundulus heteroclitus (Stage are arbitrarily designated by Armstrong and Child 1966) (a) unfertilized egg showing fibrous chorion, C and oil droplets within the yolk mass. (b) The fertilized egg, showing the protoplasmic cap. (c-c') First cleavage, Meridional. (d & e) Second cleavage. (f) Third cleavage. (g) Fourth cleavage. (h) Fifth cleavage. (i) Beginning of gastrulation. (j & K) Embryonic shield in successive stages of growth. (1) The future central nervous system becomes evident.

8. Blastula at 6 to 8 hours. Blastodisc elevated to form sub-germinal cavity and marginal syncytium; beginning of epiboly
9. Gastrulation begins when the blastodisc has grown over about one-third of the egg, accomplished by invagination and accompanied by delamination and epiboly. The blastopore is the uncovered yolk area.
10. Embryonic development:
    a) **First day:** development of embryonic shield by migration of cells, covering about 3/8 of yolk by blastodisc; optic vesicles appear; somite number about ten.
    b) **Second day:** blastodisc covers 5/8 of the yolk; heart beats 50-60 times per minute; blood is colourless; Kupffer's vesicle appears; 18 somites.

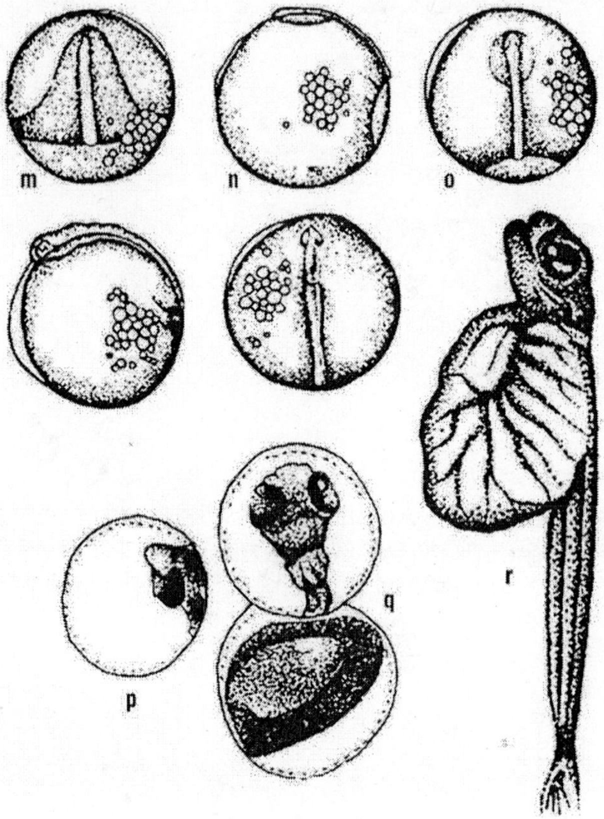

Fig. 19: Cont. (m) the emergence of the body form and closure of the blastopore. (n, o=& P) Body becomes more elevated. (q) Embryo becomes visible; embryonic movement begins; heart beats starts; blood slightly pink; numerous chromatophores hatching time newly hatched embryo.

c) **Third day:** embryo covers 3/4 of the yolk; heart beats as in adult; 150 to 170 per minute; blood slightly pink; pigment eyes; somites 20 to 28.

d) **Fourth day:** embryonic movement begins; blood is red; somites 30 to 38.

e) **Sixth day:** numerous chromatophores; hatching time (Maturity in 1 to 1½ months; life span about 1 to 2 years).

CHAPTER 20

Embryonic Development of a Frog

Study of Normal Developmental Stages of the Indian Skipper Frog, *Rana cyanophlyctis* (*Ranidae, Anura, Amphibia*).

Normal tables of successive ontogenetic stages are very useful for comparative, experimental descriptive investigations on any species during its developmental period. All embryos, in the course of their development, pass through definite chronological stages.

Amphibian eggs have long been favorite objects for investigation because they offer the research biologist a number of important advantages. In the first place, their relatively large size makes them useful, the eggs of most frogs, toads, and salamanders are big enough to be seen easily with the naked eye, and details of their development can be followed with a hand lens. They have no shells or other opaque coverings, therefore, one can readily watch them developing in a little pond water. They do not need to be maintained at any special temperature.

They develop in water, one can easily test how various chemical substances affect them merely by dissolving the substances in the water. Experimental embryologists have also found amphibian embryos useful because their tissues, though soft, are firm and rather dry, and thus are, suitable for various kinds of experiments. Finally, the eggs of some frogs and salamanders are quite easy to obtain in great numbers.

The exercise is designed with a purpose to familiarize the students with the central theme of vertebrate development from fertilized egg up to the adult. Since the Indian skipper frog, Rana cyanophlyctis is very abundant and is commonly found in almost all parts of the country, this frog species is good choice for laboratory studies of the development of frog.

Materials

Spawn (egg mass of Rana cyanophlyctis). Glass aquarium. Insect collection net, Large glass or earthen troughs of 40 to 50 cm diameter and 6 to 8 cm depth, Clean filtered pond water, Dissecting instruments, Blotting paper, 4% Formalin solution, 1:4000 solution of MS 222 (sandoz), Aqueous Bouin's fluid stereoscopic binocular microscope, Spinach leaves, Petri dishes, Specimen tubes.

Procedure

1. Visit small, shallow clear-water ponds at the time of sunrise during period of rainy season.
2. Look for the amplecting couple of male and female frogs daily till it is located.

3. Catch the couple with an insect collection net and transfer to an aquarium in the laboratory.
4. Collect the eggs as soon as they are laid and fertilized and put them in groups of about 50 eggs each in several earthen troughs filled with water.
5. Observe the eggs and embryos under binoculars to study the morphological features of various developmental stages.
6. Preserve various embryonic and larval stages of development, eggs and pre-tadpole stages in 4% formalin solution after removing the jelly capsules, and tadpole stages in Bouin's fluid after narcotising in 1:4000 solution of MS222. For removal of jelly first immerse the eggs/embryos in boiling water for about a minute and then roll on blotting paper with a pair of curved forceps.
7. Feed the tadpoles with freshly boiled spinach.

Description of Stages

The stages in the entire developmental series are selected on the basis of external morphological characteristics only the complete development of *Rana cyanophlyctis* is divided into 5 stages. The measurements made in different stages include diameter of the egg and early embryos, total length (snout to tail tip), body length (snout to base of tail), tail length (tail base to tail tip) and hind limb length.

Stage 1. Unfertilized Egg

The freshly spawned egg is spherical in shape with the animal hemisphere pigmented dark brown and the yolky vegetal hemisphere almost white. It measures about 1.5 mm in diameter. The egg is surrounded by a thin, transparent, vitelline membrane. Around the egg is the jelly capsule which swells on contact with water. (Fig. 20.1)

Stage 2. Fertilized Egg

Fertilization takes place in water. The egg (zygote) is spherical. Polar bodies may be seen near the animal pole. A faintly pigmented zone, the gray crescent is visible on dorsal side of the egg. Animal hemisphere is black and the vegetal hemisphere is whitish. As the egg enters the water the egg jelly swells and outer layers became sticky with the help of which it attaches to some aquatic plants or convenient substances. It measures about 1.5 mm in diameter. (Fig. 20.2 Age - 0 hour)

Stage 3. First Cleavage (2 cells)

A furrow appears in the animal hemisphere after fertilization. This furrow grows down through the vegetal hemisphere, dividing the egg into two blastomeres. The cleavage furrow passes through the grey crescent. The grey crescent is visible. It measures about 1.5 mm. (Fig. 20.3 Age: 45 minutes)

Stage 4. Second Cleavage (4 cells)

The second cleavage plane also commences at the animal pole, crossing the first at right angle and dividing the egg into four blastomeres. The two furrows do not meet on the animal hemisphere exactly. Two of the blastomere are in close contact at animal pole and remaining two make the close contact on vegetal pole. The grey crescent is more pronounced. It measures about 1.5 mm in diameter. (Fig. 20.4 Age-1 hour)

Stage 5. Third Cleavage (8 cells)

The third cleavage furrow is latitudinal slightly above the equator of the egg. Eight blastomeres have formed. The four pigmented blastomeres of the animal hemisphere are the smaller (Micromere) than the four unpigmented blastomeres of the vegetal hemisphere (macromeres). The size remains constant. (Fig. 20.5., Age -1.5 hours)

Stage 6. Fourth Cleavage (16 cells)

Sixteen blastomeres have formed by division of each blastomeres. The cleavage furrows are vertical. The animal and vegetal hemispheres now consist of eight micromeres and eight macromeres each. (Fig. 20.6., Age 1.25 hrs)

Stage 7. Fifth Cleavage (32 cells)

The fifth cleavage is latitudinal and asynchronous. The two cleavage furrows are formed. One in the animal hemisphere and one in the vegetal hemisphere. The cleavage furrow in the animal hemisphere divides the micromere completely and equally but the cleavage furrow in the vegetal hemisphere divides the macromeres unequally. From this stage onward increasing irregularities of the cleavage plane are more apparent. The size remains constant. (Fig. 20.7., Age - 1.45 hours)

Stage 8. Sixth Cleavage (Morula)

The furrows of the sixth cleavage are vertical and asynchronous. These furrows divide the micromere equally but again the macromere divide unequally. The cells in the pigmented area become smaller while the cells of vegetal hemisphere remain larger. (Fig. 20.8., Age - 2.45 hours)

A shallow cavity between the micromeres at the animal pole indicates the beginning of Segmentation cavity. This stage marks the beginning of blastulation.

Stage 9. Large Cell Blastula

The number of blastomeres has further increased and now along a plane across the animal pole 10-12 micromeres can be counted. Both micromeres and macromeres are recognizable individually but the former are more numerous. The surface of the embryo resembles a cluster of beads especially in the animal hemisphere. Internally the segmentation cavity or blastocoel becomes enlarged. (Fig. 20.9., Age - 3.30 hours)

Stage 10. Medium Cell Blastula

The blastomeres are more numerous and much smaller in size than before. The surface of the embryo now looks granular especially in the animal pole view. On vegetal half the macromeres now lose their individual identity. Thus morphologically vegetal half looks smooth. It measures about 1.6 mm in diameter.

Stage 11. Fine Cell Blastula

The cells are very small and surface of the embryo is almost smooth. The pigmented region extends slightly over the equator into the vegetal hemisphere indicating the beginning of epibolic movement of the micromeres onto the macromeres.

At this stage the roof of blastocoel expands and actually becomes thinner. It's margins push downward on all sides, while at the same time the floor of the yolk laden vegetal cells bulges upward from below. The blastocoel occupies the animal hemisphere and becomes larger than before. Its diameter ranges from 1.6 to 1.8 mm.

At the end of blastulation cells start to show a movement. In living condition the different blastomeres get arranged in such a way that embryo looks like a banded ball. In fact these bands or zones show the fate map stage of its development.

Stage 12. Germ Ring Stage

After the fate map formation blastomeres show a rapid division on the animal hemisphere. The resultant micromeres/pigmented blastomeres show downward movements towards the vegetal pole of the blastula. The advancing margin of rapidly dividing cells marked as a circular girdle which is known as germ ring. (Fig. 20.10., Age - 19.55 hours)

Stage 13. Initiation of Gastrulation

The micromere divides faster in the equatorial region of the blastula resulting in an extension of the micromere towards the vegetal hemisphere. The micromeres grow over the macromeres. This extension of micromeres over the macromere is called epiboly.

A distinct concentration of pigment is evident slightly below the equator on the dorsal side of the embryo. This marks the beginning of invagination in the middle of the future of the dorsal lip of blastopore. Gastrulation has now begun. (Fig. 20.11., Age - 24.55 hours)

Stage 14. Early Gastrula

The dorsal lip of blastopore has formed and is crescent shaped. The unpigmented zone of the vegetal hemisphere is reduced due to continued migration of the pigmented micromeres toward the vegetal pole. (Fig. 20.12., Age 29.55 hours)

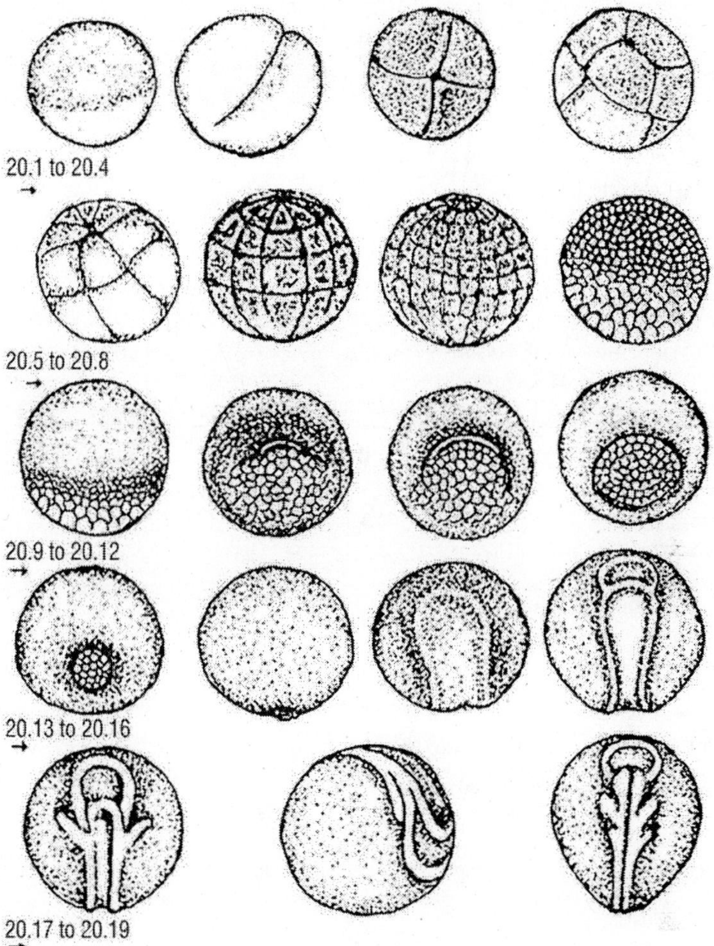

Fig. 20.1 to 20.19: Various embryonic stages of a frog.

Stage 15. Mid Gastrula

The lateral lip of blastopore have also formed which is now semicircular or horseshoe shaped. Continued epibolic migration of micromeres over the vegetal hemisphere has greatly reduced the exposed area of unpigmented yolky macromeres surrounded by the incomplete blastopore. (Fig. 20.13., Age. 32.25 hrs.)

Stage 16. Late Gastrula

The ventral lip of blastopore has made its appearance and the blastopore is now circular. The un-invaginated macromeres, surrounded by the blastoporal lips, protrude out a little and constitute the yolk plug.

In living condition it has been seen that at the dorsal lip surface cells roll inward and the invagination is seen in the macromeres of the vegetal region stream inward increasing the archenteron (Fig. 20.14., Age. 33.25 hrs.)

Stage 17. Constriction of Blastopore

Diameter of blastopore is reduced due to constriction of the blastoporal lips. The yolk plug is smaller but protrudes more to the outside than in the previous stage. The embryo is still about 1.6 mm in diameter. The initial diameter of yolk plug is about 0.89 mm. and at the time of complete closing it reduces to 0.1 mm and ultimately disappear in some of the embryos while in others it remains for longer time upto mid neurula stage. This whole process takes about eight hour.

After completion of gastrula the blastopore appears as a small, round opening filled with yolk plug. As it continues to diminish in size, it assumes a pear shaped outline through the mutual approach of its lateral lips.

Eventually these fuse completely to form a very small longitudinal groove, ending dorsally and ventrally in a small aperture. Of these two the lower one closes at once, leaving only a small depression (anal pit). The upper one remains open for a slightly longer period and closes later when the neural grooves are laid down, thus initiating the neurulation. (Fig. 20.15., Age. 38.55 hrs.)

Stage 18. Initiation of Neurula

The embryo is still spherical and is rotated to rest with the dorsal side up. The dorsal surface has flattened to form the neural (medullary plate which is outlined by pigment concentration along its edges. This indicates the beginning of neurulation. The blastopore is further constricted hut in some cases small yolk plug is still protruding through it. (Fig. 20.16., Age. 41.25 hrs.)

Stage 19. Early Neural Folds

The embryo now slightly elongated in its diameter. Its antero-posterior length ranging between 1.9-2.0 mm. The initial elevation of neural folds has clearly delimited the neural plate. The fold around the cerebral region of the neural plate is well marked hut caudally the folds gradually fade out. A median groove in the neural plate is evident in the caudal region anterior to the blastopore. (Fig. 20.17., Age. 41.55 hrs.)

Stage 20. Definitive Neural Folds

The neural folds are very prominent extending caudally to surround the slit blastopore. The broad cerebral and narrow spinal cord regions of the neural plate give it a key-hole shape. The median groove in the neural plate extends upto the cerebral region. (Fig. 20.18 Age. 42 hrs. 20 minutes.)

Stage 21. Neural Groove

The embryo is now more elongated antero-posteriorly. The elevated neural folds have grown towards each other both in the cerebral and spinal cord regions forming a groove, the neural or medullary groove. The slit blastopore is reduced to a. small pore. (Fig. 20.18 Age. 47.50 hrs.)

Stage 22. Late Neural Groove

The embryo is oval with anterior end broad and posterior end narrow. The neural folds have further converged medially particularly in the nuchal region. The neural groove in the spinal cord region is quite narrow but is still broad and shallow in. the cerebral region. Slight depression at the nuchal level on each lateral side faintly delimit the head and trunk regions. Rudiments of gill and sense plates are indicated by faintly visible elevations on each side of the head region. (Fig. 20.19., Age. 48.50 hrs.)

Stage 23. Initial Neural Tube

The embryo has slightly increased in length. The neural folds are touching each other medially along their entire lengths except anterior most region. The fusion of the neural folds begin at nuchal level and proceeds posteriorly first and then anteriorly. The rudiments of gill and sense plates are marked. (Fig. 20.19., Age. 49.20 hrs.)

Stage 24. Neural Tube

The neural folds have completely fused to form the neural tube which stands out as a mid-dorsal ridge. The median suture of the fusing neural fold is indicated by a darkly pigmented strand.

The rudiments of gills, and sense plates are better marked. In the head region there is a narrow cleft between the neural folds. At the ventral level of the hind brain the ectoderm is drawn in, forming a groove. On the ventral surface at the anterior end are two swellings the primordia of the fusion of neural folds and at the ventral end of the strip a depression, resulting from this fusion, marks the beginning of proctodaeum., Slight transverse grooves on the sides of the embryo indicate the presence of 4-6 somites. (Fig. 20.20., Age. 50.05 hrs.)

Stage 25. Tail Bud Stage

A small outgrowth, the tail buds, appears at the posterior end of the embryo. It is wider than long and is directed dorso-posteriorly. The bulges of gill plates are now very prominent infront of which, there is a small optic bulge on each side. Anterior to each optic bulge a pigmented spot marks the position of the future olfactory pit. In the mid-ventral region of the head a shallow longitudinal depression is the stomodaeal groove. Behind this groove a pair of rudiments of oral suckers appear.

In the trunk region about I pair of myotomes is distinguishable faintly. At the posterior end a slit-like depression below the tail bud indicates the formation of an anal opening proctodaeum.

After one and half an hour of initial tail bud stage the embryo begin to lengthen, particularly the tail bud grows faster. It becomes longer than wide and then its length increases fastly. Besides these features the embryo now has begun to loose its spherical form, so that the convexly curved line of the back becomes first straight and then actually concave.

The embryo by this time starts to, rotate slowly within its membranes.

In the anterior region the complete closure of the neural tube is somewhat delayed because of the greater breadth of the medullary plate at this point. However, at the time of closing there are constrictions, which divide the brain into forebrain, mid brain and hind brain.

In the head region besides the optic protuberances other sense organs also starts to develop particularly the olfactory placodes develop beneath the corresponding optic bulges. (Fig. 20.21 a-c., Age. 55.05 hrs.)

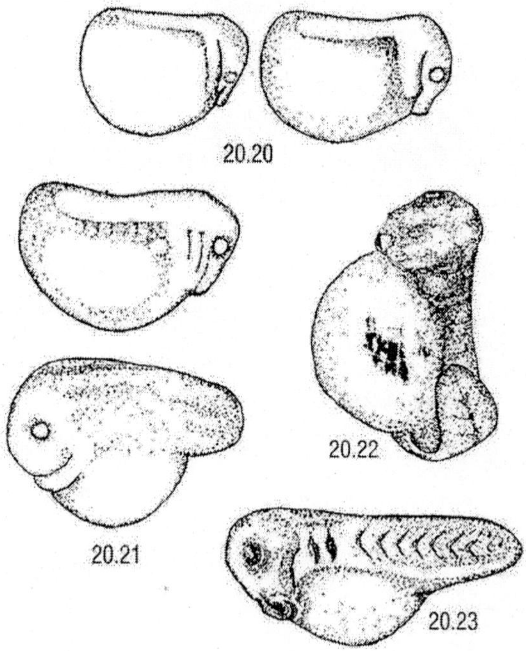

Fig. 20.20 to 20.23: Various embryonic stages of a frog.

Stage 26. Fin Formation

The tail bud is now longer than wide. The dorsal fin is marked and it extends from behind the head of the tip of the tail bud. The ventral fin is also indicated but faintly. The dorsal profile of the body is concave.

Each gill plate is divided into three prominent ridges by vertical cleft. A small pigmented depression in front of each optic bulge marks the position of olfactory pit. The stomodaeal groove has become a little deeper and anterior to this a small pit marks the position of hypophyseal invagination. Oral suckers are prominent.

In the body about 15-17 pairs of shaped myotomes extending up to the middle of the tail can be recognised. The pronephric bulge can be seen below the mytomes on each side.

The embryo, within the jelly capsule exhibit spontaneous muscular response. (Fig. 20.22., Age 59:05 hrs.)

Elevation of the Embryo

Along with the above developmental changes the embryo now raises completely above the yolk. At this stage the yolk sac looks like a sac hanging from the ventral side of the embryo. As a result of the elevation the embryo becomes first cylindrical in appearance and then it becomes "9" shaped because of curving of the tail. Rapid development of the embryo in length makes it appear as if it were riding on the yolk ball.

Stage 27. Gill Bud Stage

The tail bud is more elongated and has assumed the shape of a distinct tail and it is curved to left side. The dorsal and ventral fins are well marked, transparent and continuous around the tail tip. The myotomes increase in number and extend upto the tip of the tail.

A pair of external gill protuberance have erupted from each gill plate.

The anterior part of the stomodaeal groove has deepened to become cup shaped stomodaeum.

The oral sucker's are prominent. The optic bulges are demarcated and anterior to each bulge the olfactory pit is distinct. The beginning of the cloacal tail piece is evident at the base of the tail ventrally. The embryo is still in the jelly coat.

It shows frequent muscular movement along with the anti-clock wise movement within the jelly coat. (Fig. 20.23., Age 68.05 hrs.)

Stage 28. Early External Gill Stage

The buds of the 1st and 2nd pair of external gill have elongated. The third pair of external gill has also erupted but is very small.

The stomodaeum is pow a deep pit whose oval outer opening is the larval mouth. The oral suckers are large and nipple shaped. The olfactory pit have deepened, pronephric bulges are well recognized and myotomes extend up to the tail tip. The tail is more curved & is extending up to the third pair of gill. The number of myotomes increases. The dorsal fin approaches in the middle. At this stage the embryo shows more frequent muscular movement by the tail within the jelly coat. (Fig. 20.24., Age 98.05 hrs.)

Stage 29. Definite External Gill Stage

The three pairs of external gills are well formed & the finger like filaments become branched. Melanophores have appeared on the dorso lateral surface of the head, trunk and tail. The larval mouth consists of a pair of white beaks. It has not yet opened into the foregut. The cornea is becoming transparent and the lens is visible as a white spot. A lateral line is visible on either side of the tail in the living embryo.

At late stage the head is marked off from the trunk by a notch. The external gills become plume like and much branched. The blood circulation also starts in the gill filament. It is more prominent in first two pairs. The tail becomes longer than the body length. Intersomatic groove continue into the tail. A prominent tail fin extends completely around the tail from the vent on the ventral side to the dorsal junction with the trunk. At this stage the movements become more frequent and vigorous. Internally the stomodaeum now opened into the foregut.

Besides the above changes the yolk of the embryo reduces upto a large extent. (Fig. 20.25 a-b., Age. 134.05 hrs.)

Stage 30. Hatching

The tadpole is about 5.8 mm from snout to tail tip. The body length is about 2 mm from snout to tail base. The cloacal tail piece is 0.4 mm in length. The tail becomes longer than the body.

The head and trunk are distinctly demarcated. The three pairs of external gill are well developed with branched free ends. The larval mouth is almost circular pit. The cornea is becoming semi-transparent and lens is still not visible. The oral suckers are prominent and nipple shaped. There is beginning of circulation in the capillary loops of the external gills. It can be seen in the living embryo.

Transverse inter somatic grooves appear on the trunk and beneath this a slight ridge marks the region of the pronephric duct on the ventral side the cloacal opening is to be seen just anterior to the base of tail.

The young tadpole shows very quick movement for a few seconds thus at this stage the larva makes casual swimming movement for a short distance but it swims very actively. (Fig. 20.26 a-b. Age. 134 hrs.)

Stage 31. Early Opercular Fold

The tadpole is about 6 mm long. The body length is about 2 mm from snout to tail base. The cloacal tail piece is 0.4 mm in length.

The epidermis is becoming transparent and pigmented. The eye balls and parts of brain are visible through the skin. The mouth is still circular without any labial fringe. The beaks also does not bear the keratinized structure. The oral suckers are present. A small skin fold has begun to grow from the hyoid arch to cover the base of external gills. This is the beginning of opercular fold. The lateral line is visible on either side of the tail. The number of myotomes increases as tail grows. The opercular folds have joined together mid ventrally to form an operculum which extends a little backwards covering the proximal parts of the external gills. Now the larva swims actively for a longer period. (Fig. 20.27. Age 5 days 22 hrs.)

Stage 32. Advanced Opercular Fold

The tadpole is about 6.3 mm in length from snout to tail tip. The body length is about 2 mm from snout to tail base. The cloacal tail piece is 0.4 mm in length.

The operculum has grown posteriorly but it is not yet fused with the belly skin. Gill filaments are thin and transparent. The larval mouth does not bear the oral fringes. The larval beaks are covered by slight keratinized sheaths. The oral suckers are found. The transparency of the skin has increased and melanophores are abundant throughout the dorsal and lateral sides of the body.

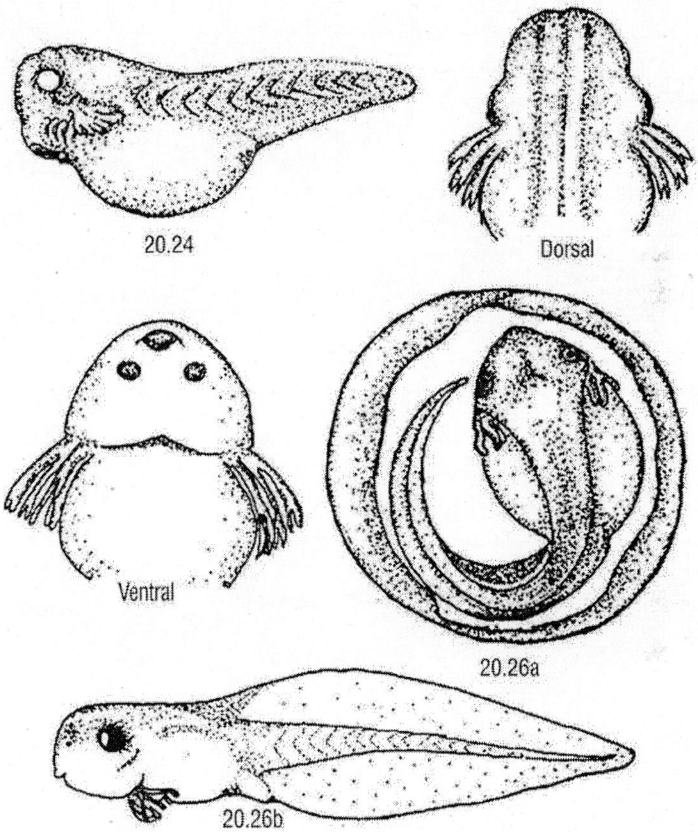

Fig. 20.24 to 20.26b: Various embryonic stages of a frog.

Myotomes are well developed in the tail region and the lateral line organs are clearly visible in le tail.

Stage 33. Operculum closed on Right Side

The tadpole is about 6.8 mm long from snout to tail tip the body length is about 2.1 mm. The cloacal tail piece is 0.4 mm in length.

The operculum has fused with the belly-skin on the right side almost concealing the external gills of this side but those of the left side are still visible. The oral suckers are found. The larval beaks are fully keratinized. (Fig. 20.28., Age. 6 days 15 hrs.)

Stage 34. Operculum Complete

The tadpole attains a total length of about 8 mm from snout to tail tip. The body length is 5 mm. The cloaca (tail piece is about 0.5 mm in length.

The operculum has fused with the skin of ventral belly wall on the left side also except in small ventro-lateral area where the tips of external gill filaments are protruding and then ultimately external gill filaments disappear completely (Fig. 12a). A hind limb bud is visible as a faint latero-ventral elevation at the junction of the trunk and tail on each side of cloacal tail piece. The oral suckers are reduced. Three rows of horny labial teeth, two pre-oral and one post-oral are now present. The mouth is surrounded by well-developed lateral conical papillae. The intestine shows 2-2½ coils. The larva is actively feeding. (Fig. 20.29., Initial hind limb bud age. 6 days 21 hrs.).

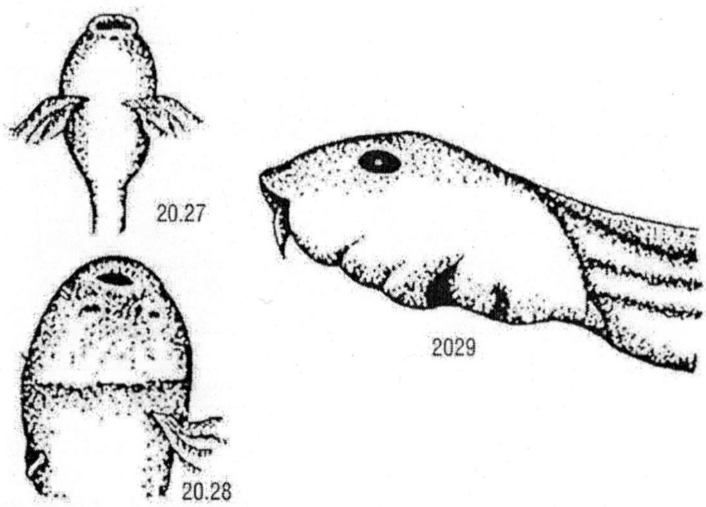

Fig. 20.27 to 20.29: Various embryonic stages of a frog.

Table. 20.1: Showing the Time, Size & Age Interval of *Rana cyanophlyctis* Reared at 20-30°

Stage Nomenclature	Time (min)	Age (hrs.)	Size (mm)
Unfertilized Egg	0	0	1.5
Fertilized Egg	0	0	1.5
First Cleavage.	45	0.45	1.5
Second Cleavage	15	1.00	1.5
Third Cleavage	15	1.15	1.5
Fourth Cleavage	10	1.25	1.5

Stage Nomenclature	Time (min)	Age (hrs.)	Size (mm)
Fifth Cleavage	20	1.45	1.5
Sixth Cleavage	60	2.45	1.5
Large Cell Blastula	45	3.30	1.5
Medium Cell Blastula	40	4.10	1.6
Fine Cell Blastula	45	4.55	1.6
Germ Ring Stage.900	19.55	1.6	
Initiation of Gastrulation	300	24.55	1.6
Early Gastrula	300	29.55	1.6
Mid Gastrula	150	32.55	1.6
Late Gastrula	30	33.25	1.6
Constriction of Blastopore	330	38.55	1.6
Initiation of Neurula	90	41.25	1.6
Early Neural Fold	30	33.25	1.6
Definitive Neural Fold	25	42.20	1.9
Neural Groove	330	47.50	1.9
Late Neural Groove	60	48.50	2.00
Initial Neural Tube	30	49.20	2.1
Neural Tube.	45	50.05	2.1
Tail Bud Stage	300	55.05	2.8
Fin Formation	240	59.05	3.4
Gill Bud formation	54b	68.05	3.8
Early External Gill Stage	1800	98.05	4.7
Definitive External Gill Stage	2160	134.05	5.8
Hatching	139		5.8

At the time of hatching the young embryo escapes from the membranous capsule. It seems to make its way out by releasing an enzyme that weakens the capsule.

Before long the rather sluggish little larva becomes an active tadpole equipped with a long tall, which it uses to propel itself rapidly through the water. During all this time the young animal has been living on the yolk that, made up the greater part of the egg. Shortly after the free-swimming stage is reached, however, the yolk is used up, and the tadpole must begin to find its own food it is obviously still a long way from being a frog. For a period of several days the little tadpole must fend for itself until it grows big enough to undergo metamorphosis and become a frog. Life is of course very hazardous for such a tiny creature without parents to provide for it. The majority of tadpoles are eaten by fish long before they reach the frog stage.

Laboratory Work

Each day review your notes from the previous laboratory session. This review will help you to see precisely what has happened to your eggs in the past 24 hours.

As you examine your embryos this week, test the responsiveness had movements of as many different stages as you can find, between the tail-bud stage and the time when the larvae can swim actively. Examine and describe the movements of each stage. Some of the things to look for and to record in your notes are the following:

a) Does the larva move just once, or more than once, when you poke it? If more than once, how many, times?
b) How would you describe the movement that the larva makes, is it bending, or rippling, or bending from one side to the other, or what?
c) Why do you think that the larva does not begin to swim as soon as it can move?
d) When do the wiggling movements cause the larva to move forward through the fluid? How far does it go? When would you say that the larva is really able to swim? How long does it continue swimming after being stimulated?

You can set an experiment on frog embryos to see the effect of temperature on their development.

Name ....

Class ....

Date ....

Effect of Temperature, on the Development of Frog Eggs

Day	Dish No	°C Dish Temp.	Stage of Embryo in Each Dish	Comments

CHAPTER 21

Post-Embryonic Development in Amphibians

Study of the basic morphological changes that take place in the external features during post embryonic stages of an anuran species (Rana cyanophlyctis) gives the knowledge metamorphic changes.

Materials

Newly, hatched tadpole larvae (of Rana cyanophlyctis), Embryos reared from batches a fertilized eggs laid in the laboratory or collected from ponds.

Binocular microscope, forceps, watch glasses, small glass dishes, cotton, drawing note book ruler or graph paper for measuring lengths MS222 (1:4000) and Bouin's solution.

Procedure

1. Examine the movements of the living tadpoles belonging to the different stages and record your observations.
2. Anaesthetize some specimens of larval and pre-metamorphic stages in MS222. Examine them one by one after laying each on a wetted pad of filter paper or cotton.
3. Examine it by the aid of the binocular microscope. Short them into definite stag numbers in separate glass dishes and label them.
4. Make a labeled drawing for each stage in one or more views so as to indicate all the characteristic features.
5. Record in a chronological order the sequence of changes that take place in the external features of the concerned stages till they complete their metamorphosis.

Observations

1. The post-embryonic distinctive features of the metamorphic stages are summarized below-
2. From "hatching" (stage-30) to "initial hind limb (stage 34) stage have been described in the chapter-Early Development of Frog"

Stage 35. Hind Limb Bud Wider than Long

The tadpole is about 14 mm long from snout to tail tip. The body length is about 4 mm. The hind limb bud is 0.5 mm long and cloacal tail piece is 0.7 mm long.

The opercular chamber opens through a short tubular spiracle on the left side.

The external gills have completely regressed. The hind limb buds are distinct and are wider than long. Two rows of pre-oral and one row of post-oral labial horny teeth are present. The intestine show 33½coils. (Fig. 21.1., Age. 7 days 18 hrs.)

Stage 36. Hind Limb Bud as Long as Wide

The tadpole is about 22 mm long from snout to tail tip. The body length is 10 mm from snout to tail base. The hind limb length is 1 mm. The length of cloacal tail piece is 1 mm.

The length of hind limb bud equals its width. Two rows of pre-oral and one row of post-oral labial horny teeth are present. The intestine s-bows 4 coils. (Fig. 21.2 Age. 13 days 18 hrs.)

Stage 37. Hind Limb Bud Longer than Broad

The tadpole is about 26 mm long from snout to tail tip. The body length is 11 mm. from snout to tail base. The length of cloacal tail piece is 1.3 mm and that of hind limbs bud is 1.7 mm.

Length of hind limb is about 1 h times its width. The bud is more or less conical, slightly constricted at the base and shorter than the cloacal tail piece. Two rows of pre-oral and one row of post-oral labial horny teeth are present. The intestine shows 4 h coils.

After about 6-8 hours the hind limb bud becomes more elongated and conical. The length being about twice its width. The bud is still without any pigment. (Fig. 21.3 Age. 22 days 18 hrs.)

Stage 38. Foot Spatula

The tadpole is about 28 mm long from snout to tail tip. The body length is 12 mm from snout to tail base. The cloacal tail piece length is 1.8 mm long. The hind limb length is 3.0 mm.

The hind limb bud is about 2½ times longer than its width. It is half of the cloacal tube in length. The distal end of hind limb bud is slightly compressed laterally to appear like a spatula. A faint constriction in the middle of the bud ventrally indicates the future knee bend. The horny labial teeth are larger in size than before hut with no further addition in the number of rows of these teeth. The intestine shows 5½ coils. (Fig. 21.4 Age 23 days 18 hrs.)

Stage 39. Foot Paddle

The total length of the tadpole is 30 mm from snout to tail tip. The body length from snout) to tail is 13 mm. The hind limb length is 4.0 mm.

Distal end of hind limb is flattened mediolaterally to form a foot paddle. A ventral depression in its middle marks the position of the knee bend which delimits the most proximal segment or thigh. The intestine shows 5½ coils. Two rows of pre-oral and one row of post-oral labial horny teeth are present. The hind limbs are shorter than the cloacal tail piece. The mouth is highly keratinized. (Fig. 21.5 Age. 24 days 18 hrs.)

Stage 40. First Interdigital Indentation

The tadpole is about 32 mm long from snout to tail tip. The body length is 13 mm from snout to tail base. The cloaca' tail piece length is 1.8 mm. The hind limb length is 4.8 mm.

A slight indentation on the margin of foot paddle on its abaxial side separates the prominences of the 4th and 5th toes. The thigh segment is distinctly demarcated. Another constriction distal to the knee-bend separates the middle segment (shank) from the distal most segment (ankle + foot] of the hind limb. Hypoaxial musculature consisting of a few transverse bands of abdominal muscles has developed underneath the belly skin. A patch of pigment cells appears for the first time on the dorso lateral part of the thigh. (Fig. 21.6 Age. 26 days 18 hrs.)

Stage 41. Second Interdigital Indentation

The tadpole is about 35 mm long from snout to tail tip. The body length is 13 mm. The hind limb length is 5.4 mm.

The margin of the foot-paddle is indented between toes 5-4 and 4-3 separating the prominence of 3rd, 4th and 5th toes. A few melanophores have appeared on the foot-paddle. The thigh, shank and ankle + foot segment are well demarcated from each other.

Stage 42. Third Interdigital Indentation

The total length of the tadpole is about 37 mm. The body length (snout to tail base) is 13 mm. The hind limb length is 7.2 mm.

The margin of the foot-paddle is indented between toes 5-4, 4-3 and 3-2 and the prominences of four toes are now separated. The melanophores make a faint discontinuous streak from the distal region of the thigh upto 4th toe. The cloacal tail piece is almost equal to the length of the limb. The intestine shows 6 h coils. (Fig. 21.7 Age. 34 days 18 hrs.)

Stage 43. Fourth Interdigital Indentation

The total length of the tadpole is 39 mm from snout to tail tip. The length of the body is 14 mm from snout to tail. The hind limb length is 9.0 mm:

The margin of the foot-paddle is also indented between toes first and second and all five toes are now separated from each other. There are distinct melanophore streaks along the 4th toe and in the indentation between the 4th and 3rd toes besides melanophore patches in the thigh, shank and ankle plus foot regions. The number or rows of the horny teeth remains constant on two pre-oral and one post-oral labial horny teeth are present. The horny beak is highoy cornified. This structure of the oral arinature persists throughout the rest of the larval life. The cloacal tube becomes short to the length of limbs. (Fig. 21.8 Age. 36 days 18 hrs.)

Stage 44. Margin of 5th Toe-Web directed towards the Tip of 2nd Toe

The total length of the tadpole is about 42 mm. The body length (snout to tail base) is 15 mm. The hind limb length is 10.5 mm.

The interdigital webs are prominent between toes 2-3, 3-4 and 4-5. Melanophores also become extensive than before cloacal tail piece extend upto base of foot. The intestine shows 7 h coils. (Fig. 21.9 Age. 37 days 18 hrs.)

Stage 45. Margin of 5th Toe-Web Directed towards the Tip of 1st Toe

The total length is about 45 mm. The body length is 17 mm. The hind limbs length is 12.6 mm.

The toes are longer and webs are prominent. The margin of 5th toe-web is directed towards the tip of 1st toe. Melonophore streaks appear on 3rd and 5th toes also. The angle between thigh and shank is obtuse. The ventral surface of each foot is dorsally applied to the ventral fin. A slight protuberance anterior to the first toe indicates the rudiment of prehallux. (Fig. 21.10 Age. 39 days 18 hrs.)

Stage 46. Margin of 5th Toe-Web directed towards Prehallux

The total length of (snout to tail tip) is about 48 mm. The body length (snout to tail base) is 19 mm. The hind limbs length is 15.0 mm.

The prehallux is very prominent. The margin of 5th toe-web is directed towards prehallux. The angle between thigh and shank is approximately a right angle. The ankle region is now demarcated from the foot region. The opening of the cloacal tail-piece is at the level of knee-bend. (Fig. 21.11 Age. 41 days 18 hrs.)

Stage 47. Demarcation between Tarsal and Metatarsal Region

The total length of (snout to tail tip) is about 50 mm. The body length (snout to tail base) is 20 mm. The hind limbs length is 16.5 mm.

The angle between thigh and shank is still approximately a right angle. The base-of cloacal tailpiece has become narrower than before. The tarsal and metatarsal regions can be separately recognized. The webs are well developed and the individual phallongs begin to be discernible even externally. The two ankles are nearly parallel of each other. (Fig. 21.12 Age. 43 days 18 hrs.)

Stage 48 Phalangeal Differentiation

The total length of (snout to tail tip) is about 55 mm. The body length (snout to tail base) is 24 mm. The hind limbs length is 8 mm.

Bases of the two thighs have come closer to each other medially reducing the proximal part of the cloacal tail-piece to a narrow strip. The thigh makes an acute angle with the shank. The ankles are parallel to or slightly diverge from each other. All phalanges are well differentiated and can be made out externally. The nasolacrimal duct is visible as a white line between the nostril and the eye. (Fig. 21.13 Age. 45 days 18 hrs.)

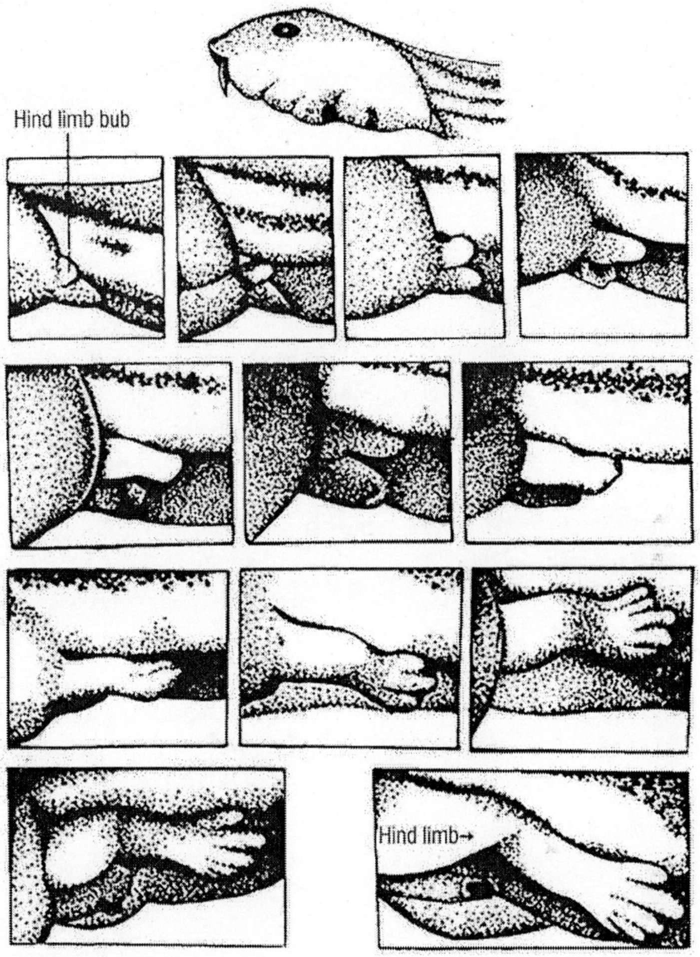

Fig. 21.1 to 21.12: Developmental stages of a frog after hatching.

Stage 49. Pelvic Symphysis

The total length of tadpole is about 55 mm. The body length (snout to tail base) is 24 mm. The hind limbs length is 20 mm.

Bases of the two thighs have almost fused with other medially indicating the formation of pelvic symphysis. Only a small distal part of the cloacal tail-piece now remains dorsal to thighs. *Linea alba*, a white ligament in the mid-ventral line is distinct. (Fig. 21.14 Age. 57 days 18 hrs.)

Stage 50. Forelimbs Visible

The total length of the tadpole (snout to tail tip) is about 55 mm. The body length remains the same as in the stage 49. The hind limb length is about 21.5 mm.

The cloacal tail-piece has almost disappeared. The forelimbs are visible through the transparent skin in ventral view and the transparent skin in ventral view and the elbows are tilting against the body wall laterally. Dissolution of the skin below the right elbow has begun to form a window for subsequent emergence of the right forelimb. The horny labial fringe start to degenerate the lower horny jaw have fallen down first. Upper horny jaw is still intact. Then even upper horny jaw have also start to fallen down. The intestine shows 9½ coils. The number of coils is at peak later on it starts to differentiate in stomach, duodenum, ileum, rectum etc. (Fig. 21.15 Age. 60 days 18 hrs.)

Stage 51. One Forelimb

The total length (snout to tail tip) of the tadpole is about 53 mm. The body length (snout to tail base) is about 24 mm. The hind limb length is about 22.5 mm.

One forelimb has emerged. The first forelimb to come out may be the left or the right. The left forelimb emerges through a window formed by enlargement of the spiracular opening. The right forelimb emerges through window formed by dissolution of the skin below its elbow. Fin resorption has begun.

After falling down of horny labial fringes the horny beak too start to degenerate. Eyes are bulging out. The nictitating membrane is indicated. (Fig. 21.16 Age. 61 days 18 hrs.)

Stage 52. (Two Forelimbs)

The total length of the tadpole from snout to tail tip is about 51 mm. The body length remain same as in the previous stage and does not change upto stage 28. The hind (limb length is about 24.0 mm. (Fig. 21.17 Age. 62 days 6 hrs.)

Both forelimbs have now emerged. The head is broader than trunk.

Table. 21.1 Showing average' total length, Body length, Tail length and Hind Limb length of the respective developmental stages of *Rana cyanophlyctis* reared at 22° to 30°C. (Means are calculated from measurements made on 5 individuals of each stage)

Developmental Stages	Total length (snout to tail base) mm	Body length (snout to tail base) mm	Tail length (tip to base) mm	Hind Limb length
Early opercular fold	6.0	2.0	4.0	
Advanced opercular fold	6.3	2.0	4.3	
Operculum closed on right side	6.8	2.1	4.7	
Operculum complete; initial hind limb bud	8.0	2.5	5.5	0.12
Hind limb bud wider than long	14.0	4.0	10.0	0.5
Hind limb bud as long as wide	22.0	10.0	12.0	1.0
Hind limb bud longer than broad	26.0	11.0	15.0	1.7
Foot spatula	28.0	12.0	16.0	3.0
Foot paddle	30.0	13.0	17.0	4.0

Developmental Stages	Total length (snout to tail base) mm	Body length (snout to tail base) mm	Tail length (tip to base) mm	Hind Limb length
First interdigital indentation	32.0	13.0	19.0	4.8
Second interdigital indentation	35.0	13.0	22.0	5.4
Third interdigital indentation	37.0	13.0	24.0	7.2
Fourth interdigital indentation	39.0	14.0	25.0	9.0
Margin of 5th toe-web directed towards the tip of 2nd toe	42.0	15.0	27.0	10.5
Margin of 5th toe-web directed towards to tip of 1st toe	45.0	17.0	28.0	12.6
Margin of 5th toe-web directed towards prehallux	48.0	19.0	29.0	15.0
Demarcation between tarsal & Metatarsal regions	50.0	20.0	30.0	16.5
Phalangeal differentiation	55.0	24.0.	31.0	18.0
Pelvic Symphysis	55.0	24.0	31.0	20.0
Fore limbs visible	55.0	24.0	31.0	21.5
One fore limb	53.0	24.0	29.0	22.5
Two fore Limbs right side	51.0	24.0	27.0	24.0
Mouth gape widening	49.0	24.0	25.0	24.0
Half tail	36.0	24.0	12.0	26.0
Small tail	30.0	24.0	6.0	27.0
Tail stub	26.0	24.0	2.0	27.5
Froglet	24.0	24.0	-	28.0

Stage 53. Mouth Gape Widening

The total length of the body from snout to tail tip of the tadpole is about 49 mm. The hind limb length is about 25 mm.

The horny beak and labial fringes have completely disappeared. The mouth gap is widening and its corners are at a level between the nostril and the eye. The tongue is present and the nictitating membrane is perceptible. The trunk is narrower than the head. About one fourth of the tail is resorbed adult type pigment patches have appeared on the dorsal surface of body and limbs. The longitudinal skin folds on the back of the trunk are conspicuus. (Fig. 21.18 Age. 64 days 6 hrs.)

Stage 54. Half Tail

The total length (snout to tail tip) of the tadpole is about 36 mm. The hind limb length is about 26 mm.

The tail has resorted to about half its length leaving a remnant of the dorsal fin. The mouth-gape is wider and its corners are at the level of the anterior margin of the eye. The snout assumes the adult shape. A faint mid-dorsal vertebral line from snout to the posterior end of the trunk is discernible. (Fig. 21.19 Age. 64 days 18 hrs.)

Stage 55. Small Tail

The total length of the tadpole from snout to tail tip is about 30 mm. The hind length is about 27 mm.

About three-fourth or more of the tail is resorbed. There is no trace of fins. Corners of the mouth are at the level of the middle of the eyes which bulge out prominently. (Fig. 21.20 Age. 65 days 18 hrs.)

Stage 56. Tail Stub

The total length of the tadpole from snout to tail tip is about 26 mm. The hind limb length is about 27.5 mm.

The tail is reduced to a stub corners of the mouth have reached the level of the posterior margin of the eyes. A tympanum is faintly visible behind each eye. The general body shape is of adult type. At this stage the froglet usually comes out of water but quite happy in the aquatic medium. (Fig. 21.21 Age. 66 days 18 hrs.)

Stage 57.

The total length of the tadpole from snout to tail tip is about 24 mm. The hind limb length is about 28 mm.

The tail has completely disappeared. A white mid-dorsal vertebral line is distinct pigment patches on sides are well marked metamorphosis is now complete.

During this process the size of the tadpole also shows drastic change. The tadpole also shows drastic change. The tadpole keeps on growing till about 2 months (at 22-30 °C in Feb.) after fertilization and then its metamorphosis begins.

The changes during metamorphosis as evidenced by emergence of forelimbs resorption and disappearance of the tail and cloaca' tube, widening of mouth opening, resorption of lips.

Along with these metamorphic changes the tadpole frequently comes to the surface of water to take gulps of air. This indicates that the lungs have started functioning. At the initial stage of metamorphosis the tadpole stops feeding. A considerable thickness of the epidermis is cast off, the horny tooth papillae and/ or horny labial fringes are removed along with the epidermis. The mouth instead of remaining circular, becomes laterally elongated. The tongue increases in size. The limbs gradually become longer and stouter. The eyes become enlarged and bulge above the level of the head. The coiled intestine shortens and the stomach and the rectum get differentiated, consequently the abdomen becomes proportionately less bulky than that of the early mature tadpole. The spiracle disappeared and the operculum becomes a part of the side of the head. The lungs take up the function of respiration. The mesonephros becomes the functional kidney. The tail is gradually absorbed. The sex becomes distinct.

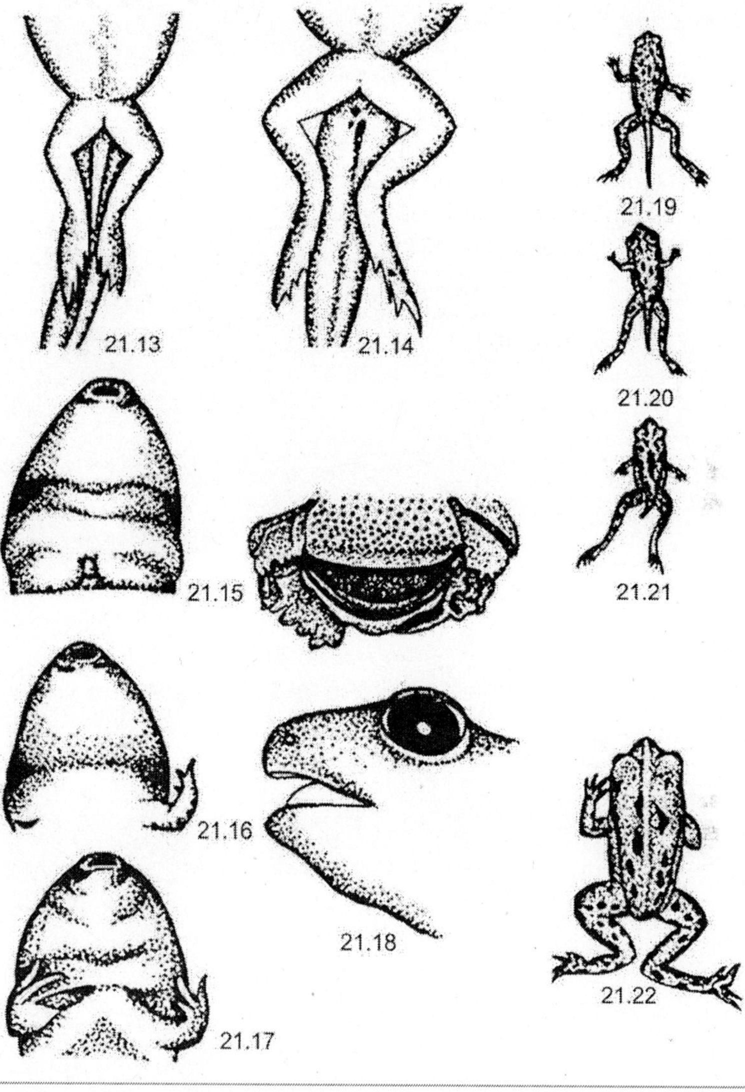

Fig. 21.13 to 21.22: Developmental stages of frog after hatching.

In this way the fish like tadpole changes into a froglet and comes frequently to the land or the surface of water to eat insects. The herbivorous tadpole becomes a carnivorous frog. The young frog becomes sexually mature after one and half year to two years. (Fig. 21.22 Age. 68 days 18 hrs.)

Table. 21.2: Age and length at some significant development stages of *Rana cyanophlyctis*.

Developmental Stages	Age	Length in mm			
		Total Body	Tail	Hind Limb	
29. Definitive External gill stage	5 days 14 hrs.	5.8	2	3.8	
32. Advanced opercular fold	6 days 9 hrs.	6.3	2	4.3	
34. Operculum Complete: L Initial hind limb bud	6 days 24 hrs	8.0	2.5	5.5	0.12
37. Hind limb bud longer than broad	22 days 18 hrs.	26.0	11.0	15.0	1.70
43. Forth interdigital indentation	36 days 18 hrs.	39.0	14.0	25.0	9.0
48. Phalangeal differentiation	45 days 18 hrs.	55.0	24.0	31.0	18.0
50. Forelimbs visible	60 days 18 hrs.	55.0	24.0	31.0	21.5
57. Froglets	68 days 18 hrs.	24.0	24.0	-	28.0

CHAPTER 22

Further Studies on Frog Embryos

Our understanding of scientific phenomena moves steadily forward for two reasons. First, the scientific method of research enables us to answer certain questions firmly and conclusively. Second, the very act of solving one question generally brings up further questions, which can then be tackled by the scientific method. It is really this second aspect of the matter that keeps science advancing steadily. In an active scientific laboratory there is never any shortage of questions.

As you work on the Block, further questions are very likely to pop up in your mind. Perhaps you would like to try to perform experiments that would answer some of your questions. This section is made up of some important problems that have presented themselves to people working in the area covered by this Block. Experiments by which these problems might be investigated are described in this section.

Series A: Further Studies on the Development of the Frog

The following questions are raised in this section-

1. Does the frog larva require practice in order to learn swimming?
2. Is the entrance of a sperm necessary to initiate the development of an egg?
3. Is the development of the egg affected by reducing the energy supply?

Experiment 1: Does the Frog Larva require Practice in Order to Learn Swimming?

If you yourself have learned to swim, you realize that it required many hours of practice in order to develop the skill. You started with simple arm and leg movements and gradually developed a smooth stroke. You will observe as you study early movements and reflexes in young frog larvae that the larvae start with simple movements, which become more and more complex until they swim quite well. Do these stages of movement indicate that the larvae must practice in order to swim, or do they simply represent stages of development of the muscles and nerves?

You may be able to answer this question by raising some embryos in a narcotic solution weak enough to allow normal structural changes to go on, but just strong enough to prevent the early movements from occurring. Your controls for this experiment will be embryos of the same stages that are allowed to develop in ordinary pond water.

First Day

Laboratory Procedure

You will need several frog larvae which have not yet reached the stage of muscular movement; MS222 solution, pond water; and culture dishes.

Place 10-15 frog larvae in a culture dish containing the concentration of chloretone assigned by the teacher. Place another 10-15 larvae of the same stage in a culture dish containing pond water. Label these dishes Narcotic solution and Control respectively and place them in the temperature gradient box or on the laboratory bench.

Second Day

Examine the embryos you placed in the narcotic solution or in pond water first day? At what stage they are? Do they move of their own accord? If not, try poking them gently with a bristle mounted on a handle. Do the controls respond to stimulation? What kind of response do they make? Do the experimentales respond? If they do, ask your teacher about substituting a stronger chloretone solution. If there is, no response from the experiment, change the solution, replacing it with the same concentration of chloretone as you used first day. Record all results, and set dishes aside for examination tomorrow.

Third Day and Subsequent Days

Examine your larvae every day. Determine whether the controls and the narcotic-treated larvae are in the same stage of development when examined. Record the kinds of movements that controls made spontaneously or in response to stimulation. Determine if the controls can move. Change the narcotic solution each day.

Continue these observations until the controls can swim at least 5 times their own length-when disturbed. How many days this will take will depend on the temperature at which the larvae are being raised, expect, or perhaps you can figure it out the effect of temperature on the development of the frog. As soon as the controls can swim well, proceed as following

Transfer the narcotic-treated tadpoles into fresh pond water. This can be done by draining off the narcotic solution and replacing it with pond water or by transferring each tadpole separately with a wide-mouth medicine dropper. Swirl or shake the dish to promote washing out of the narcotic. After 5 minutes, change the water or transfer the tadpoles again, and after five more minutes repeat the change. Stir or shake the dish frequently.

After 15 or 20 minutes the tadpoles should be coming out of their anaesthetized state. Place a dish of controls nearby and observe both groups closely. Do the narcotic-treated tadpoles begin to move? Describe the first movements they make. Do they swim? If they do begin to swim, compare their swimming with that of the controls. Is it as smooth? Is it as fast? Do they continue, swimming as long as the controls do? Record all observations in your notebook.

Experiment 2: Is the Entrance of a Sperm Necessary to Initiate the Development of an Egg?

When a ripe egg is penetrated by a sperm, two things happen (1) the egg receives a set of chromosomes that bring in the genetic characteristics of the father of the new individual; (2) the egg is stimulated to begin developing. Are these two effects inseparable from each other? Or is the act of stimulating the egg independent of the addition of a male nucleus? To try to answer these questions, we might focus our attention on the fact that the sperm must break through the surface of the egg in order to enter. Then we may ask, would breaking, through the surface by mechanical means have the effect of stimulating or activating the egg to develop, even though it contained no sperm?

We shall attempt to answer this question in the laboratory by pricking ripe frog eggs with a fine glass needle. It will help to bathe the eggs first with a mixture of frog blood and coelomic fluid.

Laboratory Procedure

You will need pond water; an ovulating female frog; glass slides; glass needles; scissors; forceps; and a pithed female frog that has been isolated from male frogs a week or more All frogs used' in this experiment must be washed in running tap water before starting the experiment.

- **Step 1.** Open the body cavity of the female using the same procedure you used to open the males. Locate the heart and clip off the tip of it as indicated by the dashed line A-A in figure. Close the skin back over the body and cover the female with a cloth dipped in standard Holtfreter's solution.
- **Step 2.** Clip the skin away from the hind leg to expose the large Muscles. Remove these muscles with scissors, and wrap them in the wet cloth.
- **Step 3.** Half fill two culture dishes with pond water and label one of them "Experimentales", and the other "Controls". Strip two ribbons of eggs from an ovulating female onto each of 5 microscope slides (30-40 eggs on each slide).
- **Step 4.** Fold back the skin of the opened frog to expose the blood-coelomic fluid mixture surrounding the internal organs. Dip the end of one of the muscles in this fluid and bathe the eggs on the slide with it. Be careful not to apply so much pressure that you injure the eggs.
- **Step 5.** Carefully prick a hole through the membrane on the dark side of each egg on three of the slides. Leave the eggs on two slides un-pricked. Place the four slides with pricked eggs in the dish labelled experimentales and carefully use a scalpel or razor blade to scrape the eggs into the fluid. Scrape the un-pricked eggs into the dish labelled "Control". Put both dishes aside until tomorrow.

Continuing this Experiment

At the next laboratory period, take out the two dishes you set aside yesterday, and, examine their contents under the binocular microscope. Bear in mind that normally fertilized eggs would be in late cleavage by now.

Do any of your pricked eggs seem to be developing? (Don't allow yourself to be fooled into thinking that degenerating eggs are developing a broken, mottled surface is simply a sign that the egg is dead). See if you can find any eggs that look like the normal development stages you have previously studied. What stages do you see? Count all the eggs you have, both dead and developing, and then discard the obviously dead eggs. Do the same for your control, un-pricked eggs. Record the total numbers of experimental and control eggs and the number of developing eggs (if any) in each group. Save any developing eggs to see how far they will go.

In case only a small percentage of pricked eggs should be activated, it will be useful if all data obtained by all classes are pooled. Your teacher will tell you how to add your findings to the findings for the entire class.

Experiment 3: Is the Development of the Egg affected by Reducing the Energy Supply?

Living processes need energy to keep them going. You know that your muscles need energy in order to contract. Many other processes, which you cannot see going on in a living system, require energy, too. For example, it takes energy to produce the enzymes that digest your food, or to pump water through your kidney tubules. This requirement for energy is just as important for embryos in their earliest stages as it is for mature organisms.

The energy that we need is ultimately derived from the oxidation or burning of the large molecules, namely, the proteins, fats and carbohydrates that we eat.

Oxygen makes energy available by participating in the oxidation (or breakdown) of large molecules, such as the foods we eat or the gasoline we use in our automobiles. When these large molecules split up, the energy which was put into building them up is released and is put to good use by living systems or by machines. When the energy from the large gasoline molecules is released in the engine of an automobile, it is used to push the pistons which= turn the crankshaft and connecting gears. These gears direct this energy to turn the wheels. When the clutch or the gears are disengaged, the energy is still being released, but it doesn't reach the wheels; thus, it is not put to use.

There are substances which will disengage the gears in a living organism, so that energy which is released cannot he put to use.

You may wish to try one of these substances and observe for yourself the results of this inhibition.

Laboratory Procedure

You will need frog embryos in the pre-gastrular stage (stage 5-7); frog embryos in the post-gastrular stage (stage 14-17); .001 (1 X 10-3) M, .005 (5 X 10-3) M and .0001 (1 X 104) M sodium axide (NaN_3); pond water and culture dishes.

Place about 5 pre-gastrular embryos in each of four dishes containing the three different concentrations of NaN_3 and pond water. Place 5 post-gastrular embryos in each of another set of solutions. Note carefully the stage of development of the embryos in each container.

Continuing this Experiment

Examine the embryos each day and record the stage of development of the embryos in each dish. When do you first notice a difference in the stage of the experimentales and the controls? Once retardation occurs, do the experimentales stop development entirely, or do they just slow down? Is retardation the same for the embryos in each concentration of NaN_3? Are both stages equally susceptible to the effect of the NaN_3? Are any of the embryos abnormal?

As soon as the embryos in a dish are found at the same stage on two consecutive days, remove half of these embryos from the NaN_3 and place them in pond water. Keep them another day. Do they resume development, or is the inhibition permanent?

CHAPTER 23

Study of Normal Stages during Chick Embryonic Development

Morphogenesis and Growth

Tile study and preparation of a series of normal stages of the chick are not only widely used in descriptive and experimental embryology but are proving to be increasingly valuable in medical research, as in work on viruses and cancer.

The egg of the hen is a very specialized type, closed box or cleidoic egg. It has no relationship with the environment except gaseous exchange. It is provided with an enormous, amount of stored food material in the form of yolk. The active cytoplasm from which the embryo is going to develop is separated out in the form of a small germ disc. It is restricted at one place over the yolk ball and covered over with a thin transparent vitelline membrane. The egg of bird develops over only a limiter-range of temperature (about 37°C-39°C), which is approximately the body temperature of the adult bird. Therefore, one should have to keep all eggs in a single incubator in which the temperature must be carefully controlled.

The total period from egg laying to hatching is 21 days in case of domestic fowl, Gallus domesticus. Though the development of any organism is a continuous process, certain stages can be observed to understand the-distinct changes that occur in the shapes of various organs.

The developmental stages of chick described by Hamburger and Hamilton (known as HH stages) are now used almost universally. The table of normal development described by Hamburger and Hamilton supersedes the earlier stage-wise description of chick embryo development. It is first necessary to go through the description of the developmental stages so as to be familiar with the diagnostic features and terminology used. See the fixed and preserve embryos arid their photographs.

Materials

Fertilized and incubated eggs. During incubation, care must be taken to rotate the eggs periodically. So the egg membranes do not stick to the shell. 0.9% Saline (about 100 ml per embryo). Dissecting instruments. Petri dishes of 10 cm diameter (1 for each embryo) and 15 cm diameter (2 for each embryo). Bowl to collect the wastes. Binocular microscope.

Procedure

1. Observe the egg and locate the position of embryo with the help of a torch light or candle light.

2. Crack a small area of egg shell at the broad end with a blunt forceps.
3. Remove the shell pieces gently, so that one can see white opaque shell membrane. There are two shell membranes. The two membranes are closely attached to each other except at the blunt end. Because of this gap at the blunt end, it is easy to open the egg from the broad end without disturbing the embryo.
4. Puncture the inner shell membrane and observe the yellow yolk ball, with the blastodisc floating in the thin albumen.
5. Remove the albumen as much as possible, with a dropper.
6. Broaden the shell opening so that entire contents of the egg can be poured in the fingerbowl containing saline.
7. Rotate the yolk ball gently and let the blastodisc be towards the observer.
8. Place a drop of 0.1 % neutral red solution on the blastodisc. After 5 minutes, carefully pour some saline on the blastodisc to wash of excess stain.
9. The red streak indicate the blastodisc.

Observations

1. Examine various areas of the embryo stained with neutral red.
2. Identify the stage of development.
3. Draw labelled diagram of the embryo.

- **Stage 1. Pre-Streak (Fig. 23.1):** Prior to the appearance of the primitive streak. An "embryonic shield" may be visible, due to the accumulation of cells towards the posterior half of the blastoderm.
- **Stage 2. Initial Streak:** ("Short-broad beginning-streak" of Spratt,) (Fig. 23.2). A rather transitory stage in which the primitive streak first appears as a short, conical thickening almost as broad as long (0.3-0.5mm in length), at the posterior border of the pellucid area. Usually obtained after 6-7 hours of incubation.
- **Stage 3. Intermediate Streak:** (10-12 hrs) (Fig. 23.3a and 23.3b). The primitive streak extends from the posterior margin to approximately the center of the pellucid area. The streak is relatively broad throughout its length, and is flared out where it touches the opaque area. No primitive groove.
- **Stage 4. Definitive Streak:** (15-16 hrs) (Fig. 23.4a and 23.4b). The primitive streak has reached its maximal length (average length = 1.88 mm,). The primitive groove, primitive pit, and Hensen's node are present. The area pellucida has become pear-shaped and the streak extends over two-thirds to three-fourths of the length.
- **Stage 5. Head-Process:** (17-18 hrs). (Fig. 23.5) The notochord of head process is visible as a rod of condensed mesoderm. Extending forward from the anterior edge of Hensen's node. The head-fold has not yet appeared. Since the length of the notochord increased during this stage.

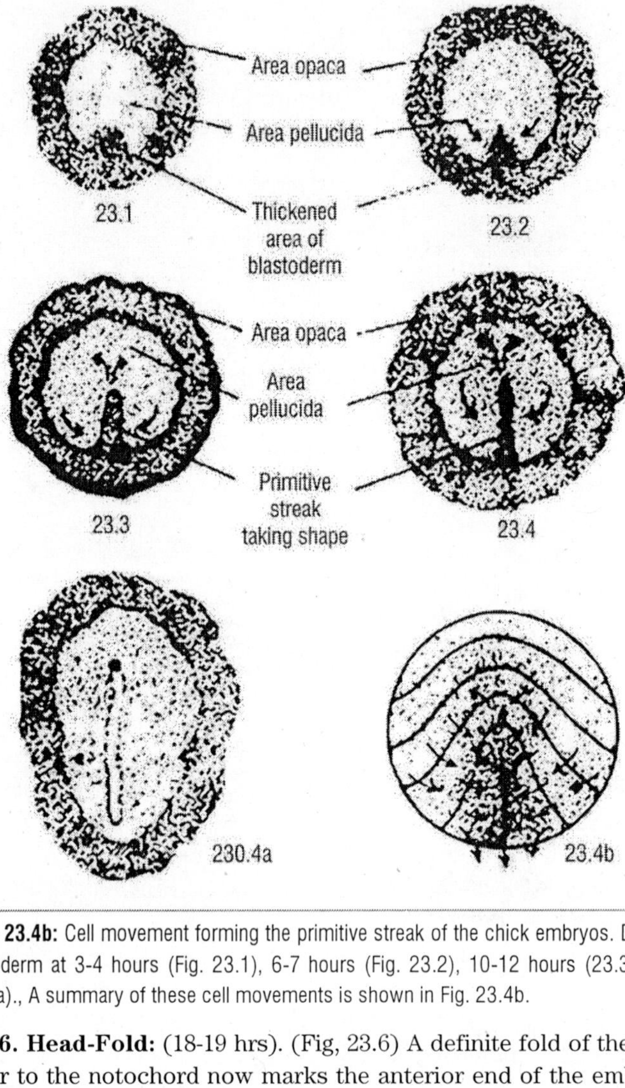

Fig. 23.1 to 23.4b: Cell movement forming the primitive streak of the chick embryos. Dorsal view of chick blastoderm at 3-4 hours (Fig. 23.1), 6-7 hours (Fig. 23.2), 10-12 hours (23.3a, b)., 15-16 hours (23.4a)., A summary of these cell movements is shown in Fig. 23.4b.

- **Stage 6. Head-Fold:** (18-19 hrs). (Fig, 23.6) A definite fold of the blastoderm anterior to the notochord now marks the anterior end of the embryo proper. No somites have yet appeared in the mesoderm lateral to the notochord. This is a transitory stage, since the head-fold and the first pair of somites develop rather closely in time

  Stages 7 to 14 are based primarily on the numbers of pairs of somites which are clearly visible. The number of somites appears to be the simplest criterion for staging this phase of development, and it is sufficiently accurate for practical purposes. A stage is assigned to every third pair of somites which is added; embryos with in-between numbers of somites are designated by adding a (+) or (-) sign to the appropriate stage. Thus, stage 7 designates an embryo with one pair of somites; Stage 7 + = two pairs; stage 8= three pairs;

stage 8 = four pairs; etc.

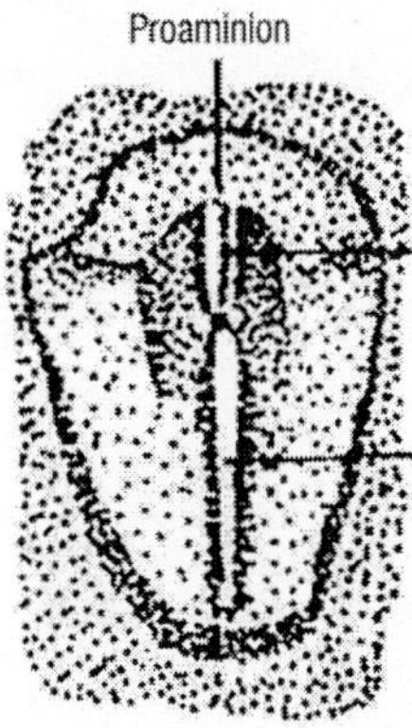

Fig. 23.5: Head process

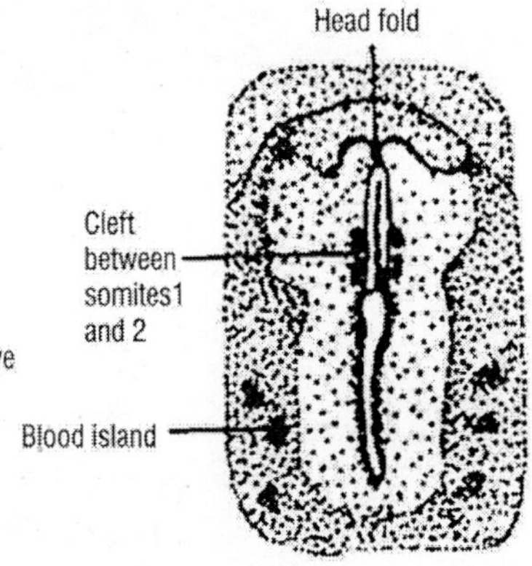

Fig. 23.6: Head fold stage.

Fig. 23.5, and 23.6: Development of primitive streak during gastrulation.

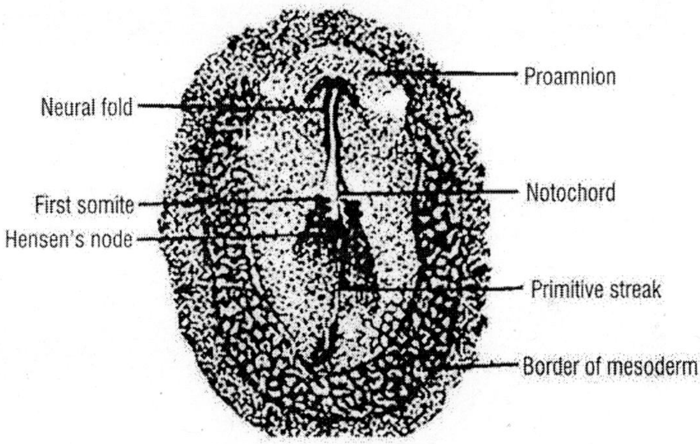

Fig. 23.7: Dorsal view of an entire chick, embryo at the beginning of somite formation. (About 21 to 22 hours of incubation.)

- **Stage 7.** One somite: (21-22 hrs.) (Fig. 23.7). This is actually the second somite of the series; number one is not yet clearly defined. Neural folds are visible in the region of the head.
- **Stage 8.** Four somites: (23-24 hrs.) (Fig. 23.8). Neural folds meet at level of midbrain. Blood-islands are present in posterior half of blastoderm.
- **Stage 9.** Seven somites: (28-29 hrs.) (Fig.23.9). Primary optic vesicles are present. Paired primordia of heart begin to fuse.

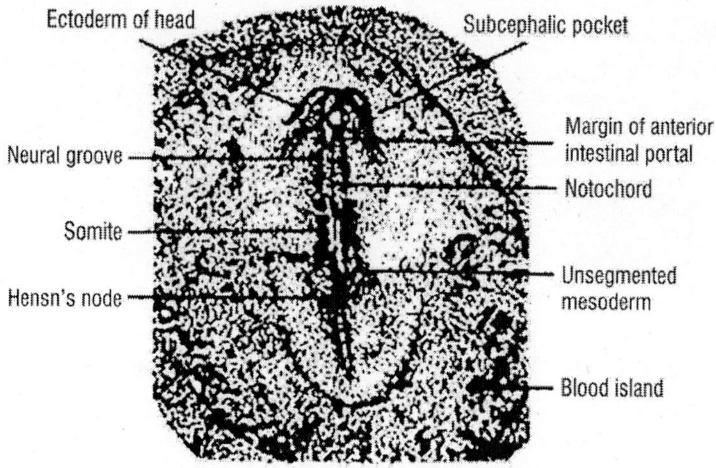

Fig. 23.8: Dorsal view of entire chick embryo having four pairs of mesoderm somites. (About 24 hours of incubation.)

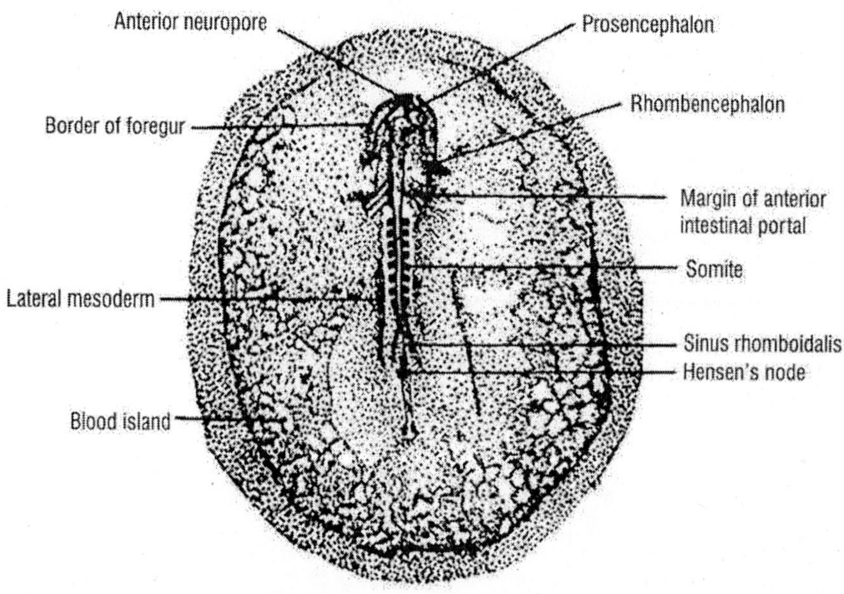

Fig. 23.9: Dorsal view of entire chick embryo having 8 pairs of somites. (About 27-28 hours of incubation).

- **Stage 10.** Ten somites: (30-31 hrs.) (Fig. 23.10). The first somite is becoming dispersed; it is not included in the counts for subsequent stages. First indication of cranial flexure. Three primary brain-vesicles are clearly visible. Optic vesicles not constricted at bases. Heart bent slightly to right.

- **Stage 11.** Thirteen somites: (33-34 hrs) (Fig. 23.11). Slight cranial flexure. Five neuromeres of hindbrain are distinct. Anterior neuropore is closing. Optic vesicles are constricted at bases. Heart bent to right.

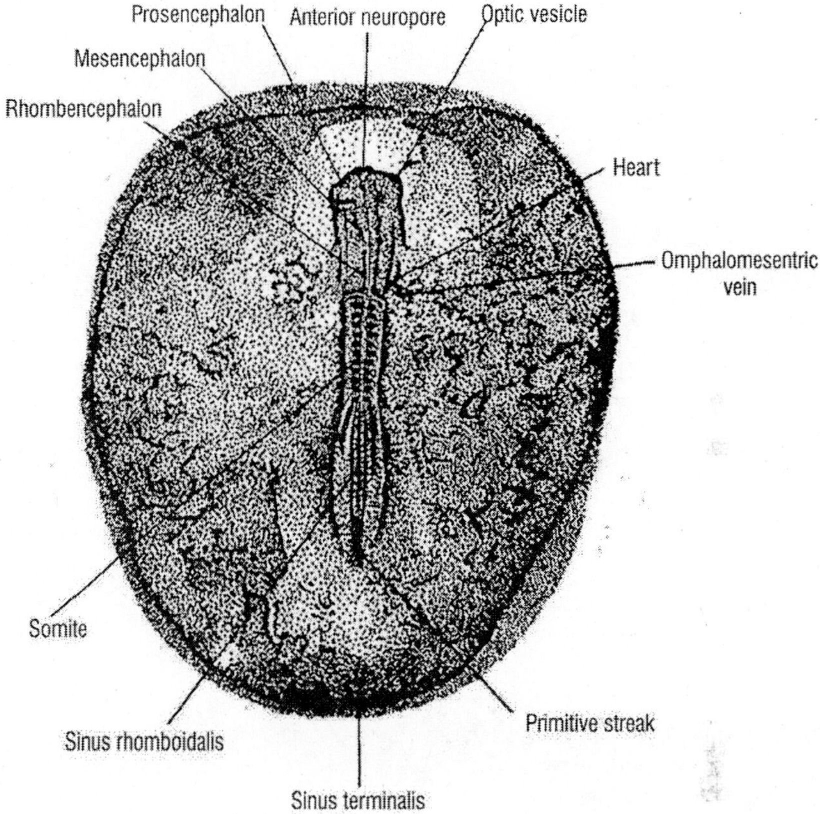

Fig 23.10: Dorsal view of an entire chick embryo having 10 pairs of somites. (About 30 hours of incubations).

- **Stage 12.** Sixteen somites: (37-40 hrs.) (Fig. 23.12). Head is turning onto left side. Anterior neuropore closed. Telencephalon indicated. Primary optic vesicles and optic stalk well established. Auditory pit is deep, but wide open. Heart is slightly S-shaped. Head, fold of amnion covers entire region of forebrain.
- **Stage 13.** Nineteen somites: (41-48 hrs.). Head is partly to fully turned to the left. Cranial and cervical flexures make broad curves. Distinct enlargement of telencephalon. Slight narrowing of opening to deep auditory pit. No indication of hypophysis. Atrio-ventricular canal indicated by constriction. Head-fold of amnion cover forebrain, midbrain, and anterior part of hindbrain.

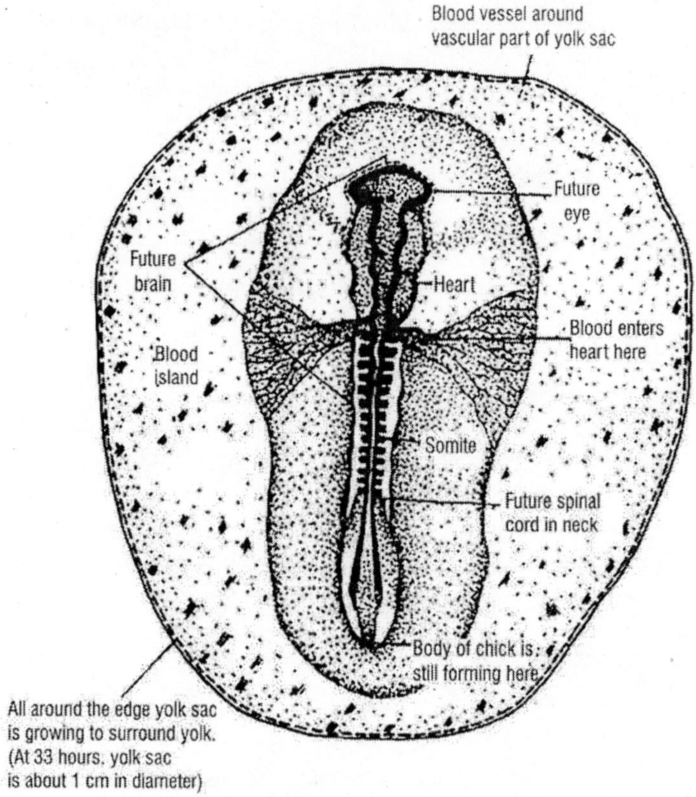

Fig 23.11: The chick embryo at 33 hours. The embryo is now spread out on the surface of the yolk, the edges of the future body being continuous with the surrounding plate of living tissue, the yolk sac, that is growing out over the yolk. On the yolk sac masses of cells are turning themselves into blood, and blood vessels are being organized and connecting with other blood vessels that lead into the embryonic heart.

- **Stage 14.** Twenty-two somites: (40-50 hrs.). Flexural, find rotation. Cranial flexure axes of forebrain and hindbrain form about a right angle. Cervical flexure a broad curve. Rotation of body back as far as somites 7-9. Behind this level, a slight flexure makes its appearance which will be referred to as "trunk-flexure".

Visceral arches 1 and 2, and clefts 1 and 2 are distinct. Posterior arches not distinct.

Primary optic vesicle begins to invaginate lens-placode is formed. Opening of auditory pit constricted. Rathke's pouch can be recognised. Ventricular loop of heart now ventral to atrioventricular canal. Amnion extends to somites 7-10.

Beyond stage 14 the number of somites becomes increasingly difficult to determine with accuracy. This is due in part to the dispersal of the mesoderm of the anterior most somites, and, in later stages, to the curvature of the tail. Total somite-counts given for the following stages are typical, but sufficiently variable so as not be diagnostic. For these reasons, the limb-buds, visceral arches, and other externally visible structures are used as identifying criteria from stage 15 onward.

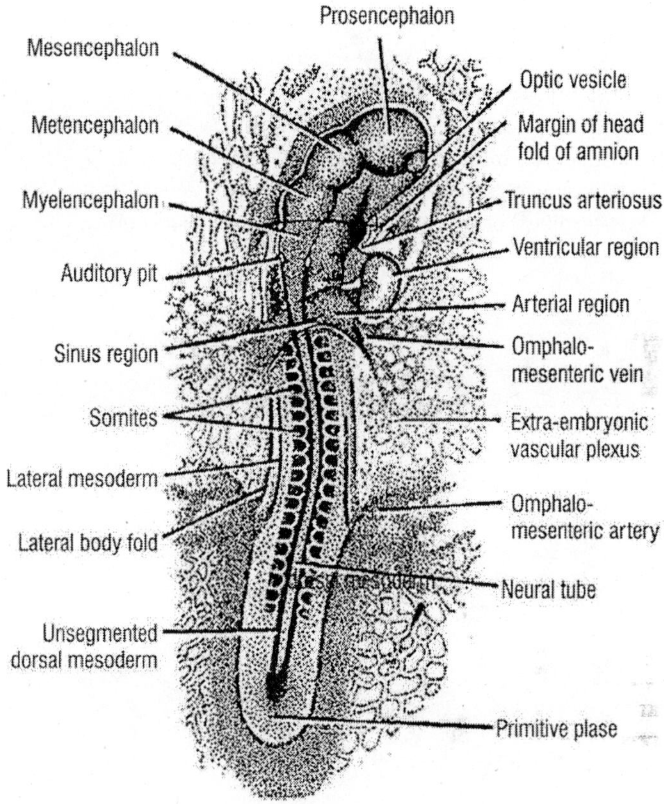

Fig. 23.12: Dorsal view of art entire chick embryo having 19 pairs of somites. (About 43 hours of incubation).

Stage 15. (50-51 hrs.). (Fig. 23.13)

1. Lateral body-folds extend to anterior end of wing-level (somites 15-17).
2. **Limb-primordia:** prospective limb-areas flat, not yet demarcated. Inconspicuous condensation of mesoderm in wing-level.
3. **Somites:** 24-27.
4. Amnion extends to somites 7-14.
5. **Flexures and rotation.** Cranial flexure: axes of forebrain and hindbrain form an acute angle. The ventral contours of forebrain and hindbrain are nearly parallel. Cervical flexure a broad Curve. The trunk is distinct. Rotation

extends to somites 11 to 13.

6. **Visceral arches:** Visceral arch 3 and cleft 3 are distinct. The latter is shorter than cleft 2 and usually oval in shape.
7. **Eye:** Optic cup is completely formed; double contour distinct in region of iris.

- **Stage 16.** About 52 hrs (Fig. 23.13). This is an embryo that was made transparent and then stained for detailed study under the microscope. At this stage, the anterior end of the body is resting on its left side. Identify the eye, brain, heart, and somites. The skirt like effect around the trunk is the posterior edge of the amnion, which is now being formed.

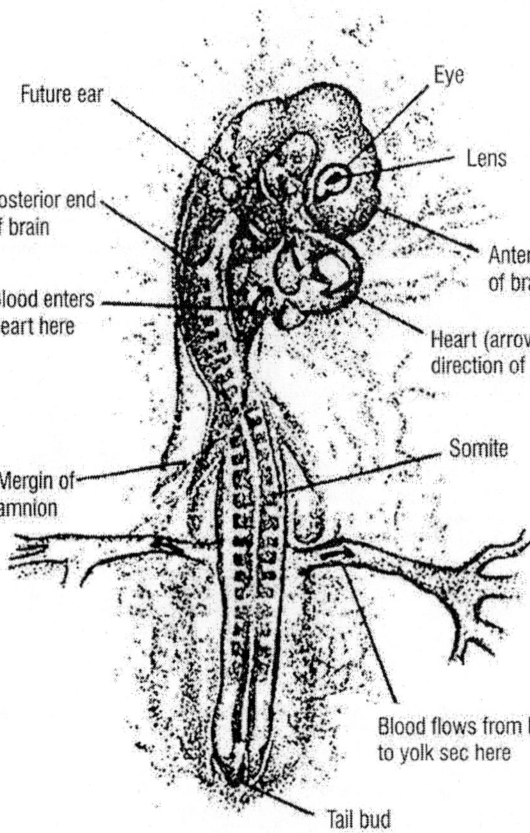

Fig. 23.13: The Chick embryo at. 51 hours. This stage shows several important changes that occur between 33 and 51 hours. The body of the embryo is now completed, and a tail is beginning to appear. The brain has bent around against itself, and the embryo is beginning to turn so that the anterior part is lying on its left side. A few hours later, the rest of the body will have rotated. The heart has been beating, at this stage and there is a lively circulation of blood through the body and the network of blood vessels on the yolk sac. The eye is further advanced than at 33 hours. The amnion, which will soon envelop the entire body, has covered the anterior end and is pushing on backward.

- **Stage 17.** About 60 hrs. The bending of the head against the trunk is continuing. The amnion by now covers the body completely.
- **Stage 18.** About 72 hrs (Fig. 23.14). The bending of the head is further advanced. The entire body now rests on its left side. The hind limb buds are visible as thickened masses of tissue on each side of the body, just in front of the curled tail tip.

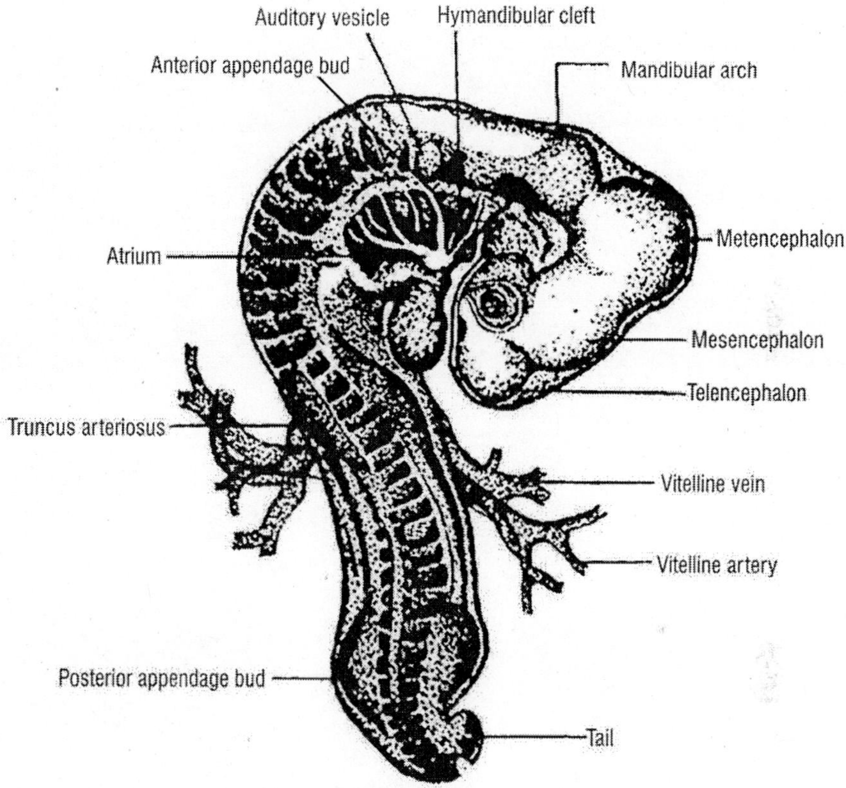

Fig. 23.14: Dorsal view of entire chick embryo. (About 72 hours of incubation).

- **Stage 19.** Between 3 and 3½ days. (Fig. 23.15) The tail tip has curled more than-before. The dark mass in front of the tail is the allantois, which is just beginning to bulge out from the body.
- **Stage 21.** About 3½ days. This is an embryo just removed from the shell and photographed by reflected light; all extra-embryonic membranes have been cut off. The body has now curved so much that the head and tail have come close together. Note how prominent the wing and leg buds have become.
- **Stage 25.** About 4½ to 5 days (Fig. 23.16). Notice the large size of the head in relation to the rest of the body. The eye can be readily identified. The limb buds are developing elbow and knee joints. Atrium

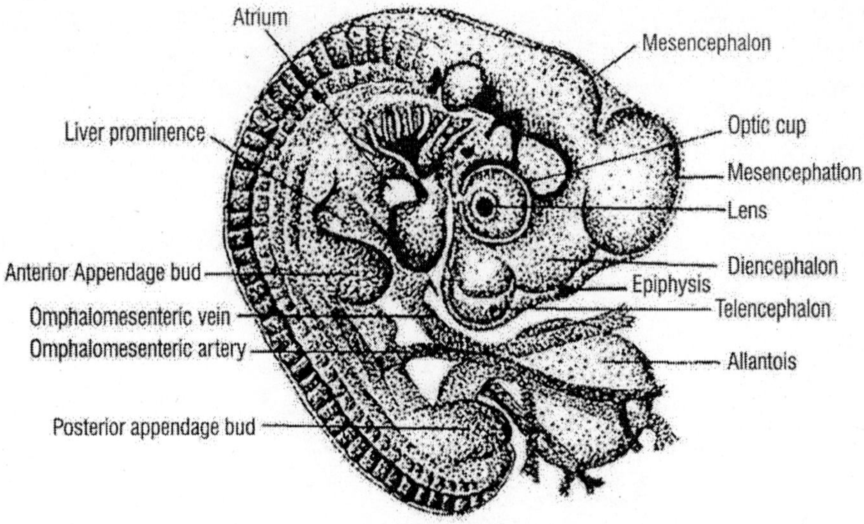

Fig. 23.15: Dorsal view of entire chick embryo. About 3½ days of incubation. Stained and cleared preparation drawn by transmitted light.

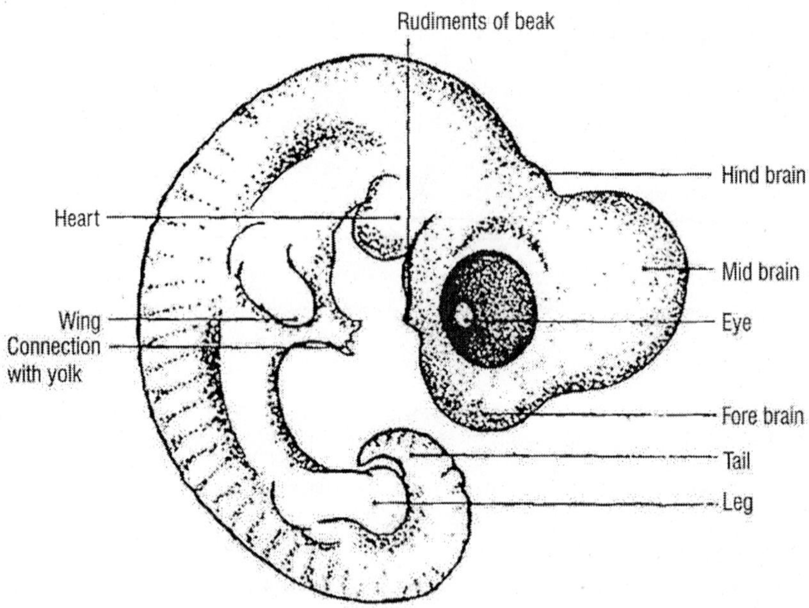

Fig. 23.16: Dorsal view of entire chick embryo. About 5 days of incubation.

- **Stage 30.** About 7 days (Fig. 23.17). The beak is beginning to appear near the lower edge of the head. A neck and trunk can now be clearly distinguished from each other. Toes are beginning to be marked out on the feet.

- **Stage 32.** About 7½ days (Fig. 23.18). The beak is now very distinct, as are the toes. Note that fingers are appearing on the wing bud, even though the chick does not have any fingers in later life.

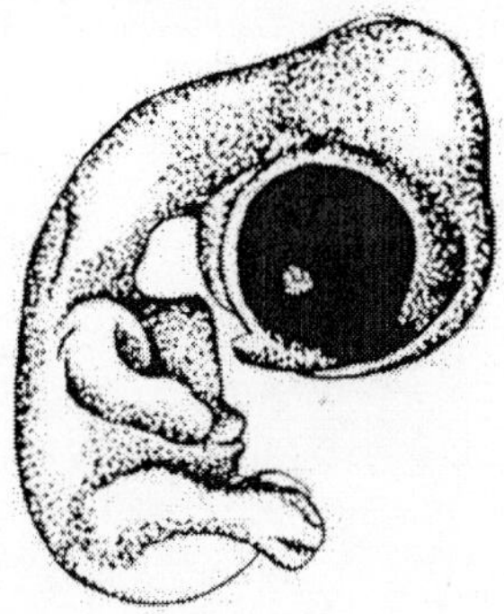

Fig. 23.17: Whole mount: 7 days old chick embryo.

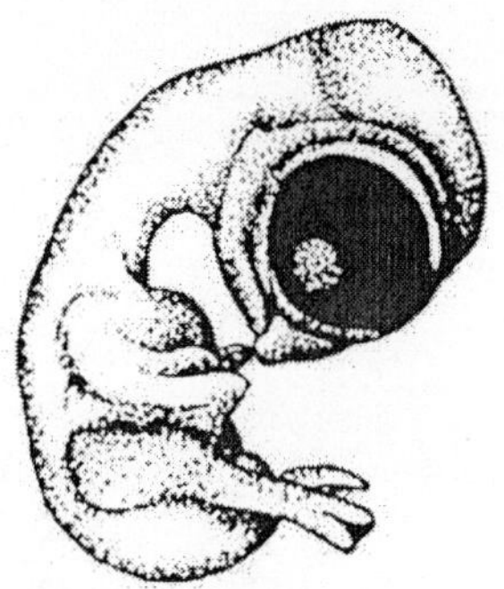

Fig. 23.18: Whole mount 7½ days old chick embryo

- **Stage 34.** About 8 days (Fig. 23.19). The small opening behind the eye is the ear. The index finger (no thumb is ever formed) is quite conspicuus.

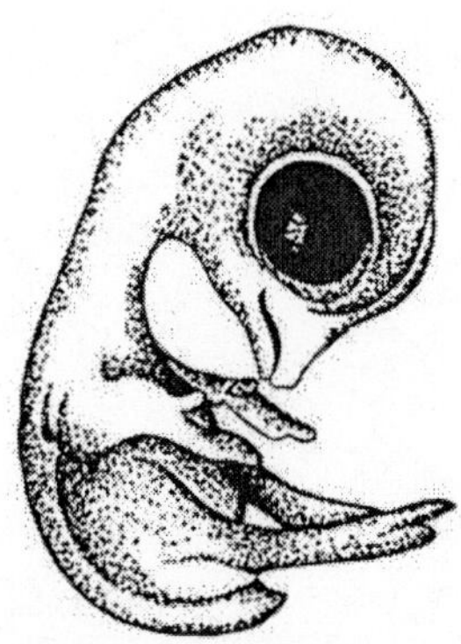

Fig. 23.19: Whole mount: 8 days old chick embryo.

- **Stage 37.** About 11 days (Fig. 23.20). Eyelids are now developing to cover the eyes. The legs are quite perfectly formed, even having claws on the toes. The skin is covered by= prominent feather buds (some of these are present as early as 6½ days, but are not large enough to show up on the photographs).

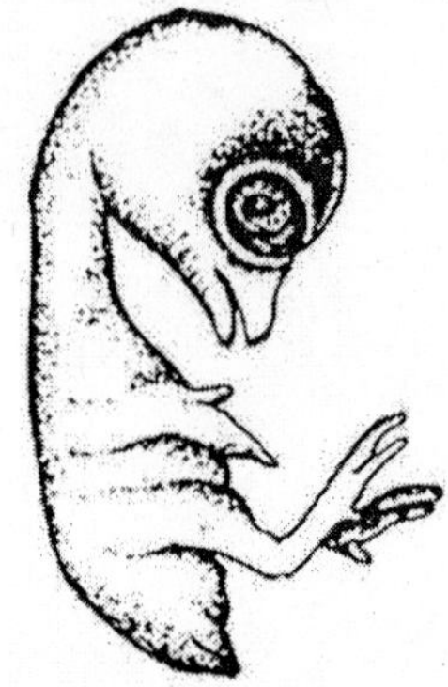

Fig. 23.20: Whole mount 11 days old chick embryo

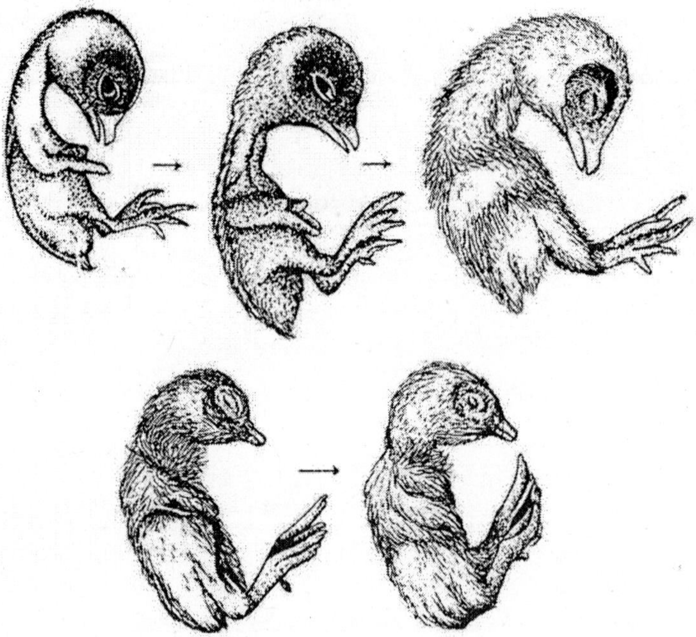

Fig. 23.21 to 23.24: Whole mounts: Fig. 23.21a,b.(12 days old chick embryo); Fig. 23.22. (13 days old chick embryo); Fig. 23.23. (14 days old chick embryo) and Fig. 23.24. (16 days old chick embryo).

- **Stage 38.** About 12 days (Fig. 23.21a and 23.21b). Lengthening of the leather buds gives the embryos a more bird like appearance. But note that the head is still very large in relation to the trunk

- **Stage 39.** About 13 days (Fig. 23.22). The feathers are now well developed.
- **Stage 40.** About 14 days (Fig. 23.23). The eyelids are now big enough to cover the eyes completely.
- **Stage 42.** About 16 days (Fig. 23.24). As the chick grows, the body is increasing "m size in relation to the
- **Stage 44.** About 18 days. The body proportions are becoming steadily more normal. Compare this stage with Stage 40 (14 days), and then contrast this 14-18 day interval with the interval between 3-7 days (Stages 18 and 30). Evidently changes occur much more rapidly early in development than in the later stages.
- **Stage 45.** About 19-20 days (Fig. 23.25). The chick has by now grown so much that it almost completely fills the shell. It is now ready to hatch. In preparation for hatching, the yolk sac has been drawn about half-way into the abdominal cavity.

Fig. 23.25: Whole mount: 20 days old chick embryo.

Process of Hatching (Fig. 23.26a, b and c)

On the 21st day the movements of chick embryo within the shell cause the head to be jerked forward and the beak upwards. The egg tooth on the beak is now quite large and these movements split the membranes inside the shell and finally crack the shell itself. The chick kicks with its legs and the beak enlarges the hole in the shell. When it first appears the chick looks rather shaggy and damp, but its downy feathers soon dry and it is now a fluffy chicken. One day old chick covered with feathers and able to peck its own food.

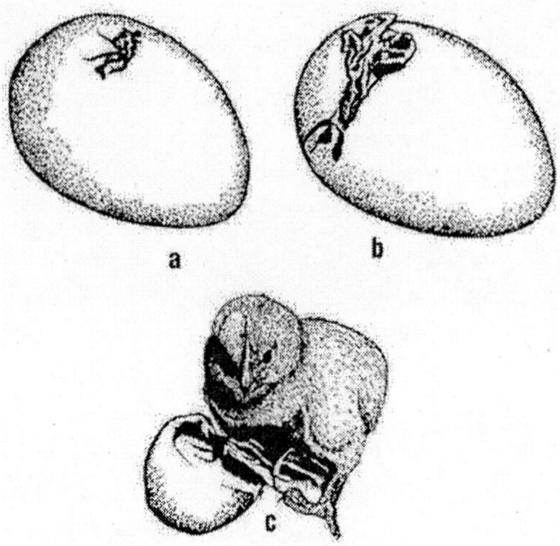

Fig. 23.26a, b and c: Process of hatching

CHAPTER 24

Further Studies on the Chick Embryo

This section describes experiments by which we could try to answer the following questions:

1. How does temperature change affect the heart of the chick embryo?
2. Does the enzyme cytochrome oxidase occur at early embryonic stages?
3. What effect does thiourea have on the thyroid glands of the chick embryo?

Experiment 1: How Does Temperature affect the Heart of the Chick Embryo?

Most warm-blooded animals (birds and mammals) maintain a constant body temperature at all times. They are able to do this by a complex series or adjustments that cause more heat to be produced when it is cold, or allow heat to be lost when it is hot. Two familiar adjustments of this kind are shivering (which is involuntary muscular exercise) and sweating. The ability to regulate temperature develops only after an animal is born or hatched and this is why young babies must be carefully protected from cold. Embryos cannot regulate their temperature at all The chick embryo, unlike the frog, cannot develop faster or slower, depending on the temperature of the air around a developing chicken egg must be kept at a nearly constant level either by the body of the hen, or by an incubator.

But now you may recall that when you studied chick embryos, you noticed that the rate of the heartbeat seemed to be slow and sluggish at low temperatures, but was more vigorous when the preparation was war. Was this just a chance variation in an embryo that was probably dying? or does the heart rate vary in some regular fashion with changing temperature? Let us set up an experiment to investigate this problem systematically.

Laboratory Procedure

You will need a hen egg incubated 60-72 hours; scissors; 4 culture dishes, one of which will fit inside the other 3; physiological saline solution warmed to 38° C; iced tap water (80°-90° C); forceps; thermometers; bright light; and a watch with sweep second hand.

- **Step 1.** Half fill the small culture dish with 38°C physiological saline; then half fill the other dishes-one with ice water, one with hot water, and one with water at room temperature.
- **Step 2.** Open an egg that has been in the incubator 60-72 hours then place the egg in the warm saline and slowly rotate the shell until the opening is downward. Remove the shell carefully to avoid breaking the yolk.

(Alternatively, one might leave the egg contents inside the opened shell, and partially submerge this preparation in water or saline at different temperatures. The temperature can then be determined inside the shell).

- **Step 3.** Use a centigrade thermometer to determine the temperature of the saline solution (or egg). Note the time and count the number of heart beats per minute. Record the number of heartbeats and the temperature in your notes.
- **Step 4.** Now, place the dish of saline (or the egg in its partial shell) in the dish of hot water and allow the temperature to climb to 40° C (not over 43° C); then remove the dish from the hot water. Note the time and plot the temperature of the saline and the number of heart beats per minute on the graph in figure 24.1
- **Step 5.** Place your dish of saline (or the egg in its partial shell) in water at room temperature. Determine the temperature inside the dish or shell, and count the heart rate (i.e., number of beats per minute) whenever the temperature drops 3° C. (For example, determine the rate, at 39°, 36°, 33°, etc.) Record the temperature and heart rate in your notebook and on the graph.
- **Step 6.** When the temperature of the preparation has dropped to room temperature, transfer into a dish of ice water. As before, determine temperature and count the heart rate at intervals of 3°. Continue until the rate falls to about 20-30 beats per minute. Then reverse the steps you have just taken, and determine the heart rate as the temperature rises back to 400-42° C. Plot your data on the graph; this dine, use a different colour or different symbol from the one you used to record rate at falling temperatures.

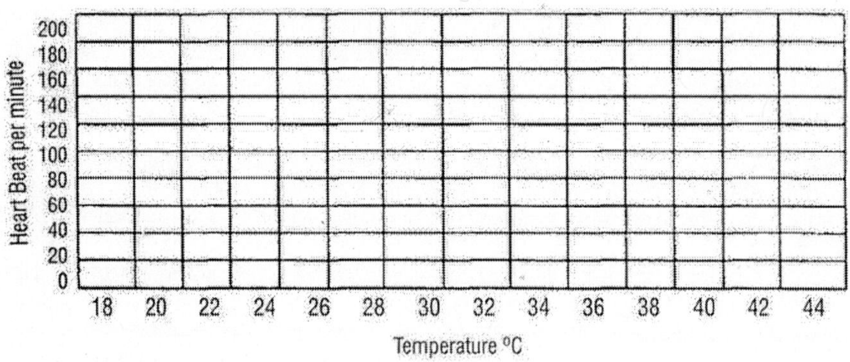

Fig. 24.1: The effect of temperature on the embryonic chick heart.

Experiment 2a: Does the Enzyme Cytochrome Oxidase Occur at Early Embryonic Stages?

The state of being alive, it is now generally agreed, depends on the organized activities of the substances we call enzymes. Each enzyme occurs in only very minute quantity, but our bodies contain thousands of different enzymes. It has demonstrated that virtually every tiny chemical change that takes place within us has a specific enzyme to carry it out.

A good enzyme to study would be cytochrome oxidase, a very important enzyme that is present in all kinds of living things. Without cytochrome oxidase the tissues all over our bodies could not make use of the oxygen our lungs take in. The deadliness of carbon monoxide and cyanide is explained by the fact that these substances stop cytochrome oxidase from working. Is this enzyme, which is a complicated chemical molecule, already present in very young embryos? Or does it only appear after some time, as the limbs or the lungs or the feathers do?

Whether cytochrome oxidase is present in small bits of tissue may be determined by the use of the Nadi technique. In using this technique, one puts the tissue to be tested the embryo in this case into a reaction mixture containing two chemical substances, para phenylenediamine (PPD) and alpha-naphthol buffered at a slightly acid pH. If cytochrome oxidase is active in the tissue, it will act on the PPD molecules in such a way as to cause them to become linked to the, alpha-naphthol molecules. The compound thus formed is bright blue, although the two substances that make it up are colourless by themselves. Therefore, if a piece of living tissue turns blue when placed in Nadi reagent, we may conclude that it probably contains cytochrome oxidase.

Laboratory Procedure

You will need an egg incubated 60-72 hours; fine, sharp-pointed scissors; filter paper; physiological saline warmed to about 38°C; petridish or Syracuse watch glass; hand lens or dissecting microscope; bright light; fine, sharp-pointed forceps; cotton wool; culture dish; paper towels; Nadi indicator solution (see appendix) and a watch.

Open your egg by the method previously used for early stages and carefully remove the embryo by the paper ring method. Place the embryo in a Syracuse dish containing some warm saline; if excess yolk makes the solution cloudy, draw it off with a medicine dropper and gently add fresh warm saline, repeating this operation until the solution is clear. Then withdraw the saline and add warm, freshly mixed Nadi solution. Record the time. Place your preparation on the stage of a dissecting microscope, against a light background, or observe it with a hand lens.

After 3 minutes, look over the preparation carefully and see if any blue colour has appeared in the body of the embryo. If so, is the colour uniformly diffused or is only a part of the body coloured? If the latter happens, encircle the affected area on figure II, and mark the time in circle. Repeat your observation every 3 minutes for a total of about 20 minutes from time.

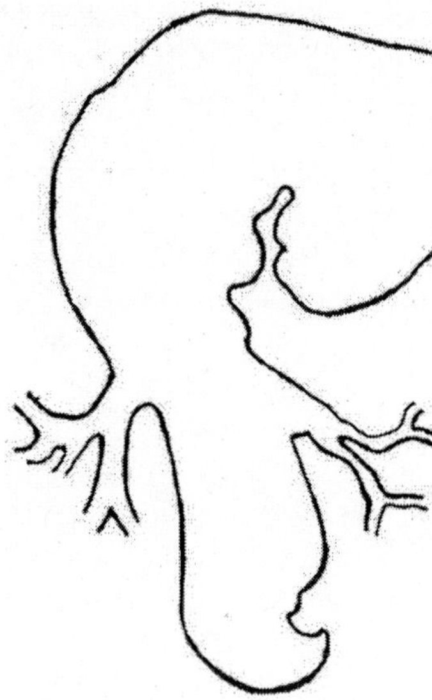

Fig. 24.2: Outline of 72-hour chick embryo

Experiment 2b: The Inhibition of Cytochrome Oxidase

If you obtained positive results in the preceding experiment, you may wonder if one can be sure that the blue colour results from cytochrome oxidase action. Perhaps you suspect that some other enzyme might have the same effect on the Nadi reagent. (After all, there are thousands of enzymes). One way to answer this question is by using an inhibitory substance known to interfere specifically with the action of cytochrome Oxidase. In a research laboratory in which adequate precautions could be taken, one would use carbon monoxide or cyanide as the inhibitor. But this substance is poisonous, we can use a safer substance, sodium azide (NaN_3). If azide prevents the appearance of a blue colour, this is good evidence that the colour did indeed result from the activity of cytochrome oxidase.

Laboratory Procedure

You will need for this experiment the same materials as for the preceding experiment, plus some warm 0.005M (5 X 10-3) M sodium azide solution in acidified saline (see appendix), or warm acidified saline for the controls.

Two squads should work together on this experiment, one squad running a control and the other running an experiment with sodium azide. The two teams should open their eggs at the same time. When the' embryos have been removed and rinsed, one squad will place' its embryo in acidified saline solution and the other will place its preparation in 0.005M azide solution. The embryos are left in these solutions for 3 minutes.

After 3 minutes the solutions are drawn off and freshly mixed Nadi reagent is poured into the dishes. Observations are then made at 3-minute intervals, as before. Record carefully the appearance of blue colour in each embryo. What kind of differences do you notice? Does the blue colour appear at the same places in both embryos? Does it take the same length of time to appear in both? Is it equally intense in both? Record all observations in your notebook.

Experiment 3: What effect does Thiourea have on the Thyroid Glands of the Chick Embryo? Or The Influence of a Thyroid Inhibitor on the Development of the Chick Embryo

The fact that the metamorphosis of the frog tadpole depends on the activity of the thyroid gland has been known for more than half a century. One might wonder, however, whether. IN activity of the thyroid gland is essential in the development of animals that do not undergo metamorphosis. For example, one might ask if the normal development of the chick embryo is in any way dependent on a supply of thyroid hormone. If the thyroid were removed, would the rate of embryonic development be slowed down? Would the body develop normally? Would the chick hatch?

It does not seem to be possible to remove the thyroid glands surgically from the chick embryo, but we can obtain about the same effect by using a chemical substance that arrests the activity of the thyroid without much affecting anything else in the body. Previous studies of the chick embryo have suggested that by 11 days the gland is capable of secreting thyroxin. At that stage the internal structure of the thyroid is similar to that in mature chickens; in addition, the gland is then able to accumulate radioactive iodine. These facts are indications of possible activity, but they do not mean that the gland is actually secreting, or that thyroxin is needed for the normal course of development. Nevertheless, the interval between 10 and 12 days would be a logical place to start if we wish to determine whether the thyroid hormone plays any part in the normal development of the chick embryo.

The chemical substance that we shall use to try to inhibit the thyroid is called thiourea. At least in mature mammals, this substance has been shown to prevent the thyroid from manufacturing thyroxin. We shall inject our 11-day embryos with thiourea today and open some of them after one week, at 18 days. The others will be kept until hatching. Should they fail to hatch, you will open them later to see what has happened to the embryos.

Schedule

(Note that the days listed in this schedule are not consecutive days, but are separated by intervals of several days.)

- ***1st day:*** Inject eggs that have been incubated for 11 days.
- ***2nd day (7 days after 1st day):*** Open and examine injected eggs (embryos now 18 days old)

- ***3rd day to 6th day (9 to 12 days after 1st day):*** Examine and weigh hatched chicks.
- ***7th day (13 days after 1st day):*** Open and examine any eggs that have not hatched.
- ***8th day:*** Discuss effects of thiourea on the chick embryo.

Laboratory Procedure

You will need eggs incubated 11 days; sterile hypodermic syringe and needle; vials containing sterile 0.25 M thiourea or sterile physiological saline solution; plastic tape; dissecting needle; absorbent cotton; small dishes; 70% alcohol. (***Note:*** When your syringe and dissecting needle are not in use, they should be kept standing in a beaker of 70% alcohol to keep them sterile. Do not lay them on the table, and do not touch the needle tips with your fingers).

- **Step 1.** Take your vial and swab off the rubber cap with alcohol. Carefully insert the needle tip through the cap. Hold the vial inverted with hypodermic pointed upward into it. Be sure fluid covers the needle tip, then cautiously pull the plunger back and draw fluid into the syringe. If nothing will flow into the syringe, inject a little air into the vial and try again. If any air bubbles appear in the syringe itself, while you are filling it, allow them to rise to the base of the needle (which is now pointing up) and push the air back into the vial. Adjust the volume of fluid so that you have 0.25 ml in the syringe barrel, then withdraw the needle from the vial and make sure there is no leakage. When you can do this easily and reliably, you are ready to inject eggs.

  If you have used a practice solution in this step, rinse the syringe 5 or 6 times in 70% alcohol, then stand it in a beaker of alcohol until you are ready to use it again.

- **Step 2.** Take one of your eggs and write on the shell C (for control) in pencil and add the number of your squad. Rinse your syringe carefully in 70% alcohol. Then, following the instructions in figure, inject the egg with 0.25 ml of physiological saline. Be careful to insert only the tip of the needle into the egg. Cover the opening with tape, and return the egg to the incubator. Rinse syringe at least 5 times in alcohol, then prepare a second egg and inject it with 0.25 ml of 0.25 M thiourea solution. Mark this egg "TU". Return egg to the incubator, then inject a third egg in the same' way with thiourea. These injected eggs should not be turned, since daily turning is necessary only in early stages of development.

  2nd Day: Open some of the eggs injected with saline or thiourea one week ago. Get out your preserved 10-11 and 14-15 day embryos for possible comparison with thiourea-injected embryos.

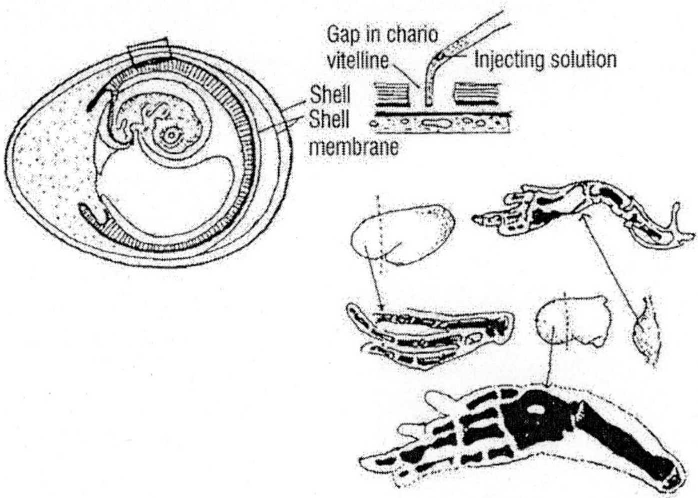

Fig. 24.3: Procedure for injecting solution into chicken eggs.

Laboratory Procedure

You will need scissors and forceps; glass dishes; warm physiological saline solution; specimen jar; formalin or 70% alcohol.

- **Step 1.** The embryos are now 18 days old. Take the egg and open it. If your squad has a saline-injected control, work in collaboration with a squad having a thiourea-injected egg, or vice versa. Thus you will be able to make comparisons as you go along. (***Note:*** If any dark fluid has run out of, the injection hole, or if the egg gives off a bad odour on being cracked, wrap it up and discard it. It has been infected by careless injection technique).

  First examine, the chorioallantois. Does it look normal in the thiourea-treated specimen? Record its condition. Then tear away the chorioallantois so that you can study the chick embryo, amnion, and yolk sac.

  Amnion. Is it present in the thiourea-injected specimen? Is it complete? Make notes on what you' see.

  Chick embryo. Is it the same size in control and experimental eggs? Do you see any abnormalities in the experimental embryo? Comparing it with your preserved 10-11 day embryo (i.e., taken at same stage as the injection was made), does it appear that growth continues after the administration of thiourea? Does each chick have an egg tooth?

  Yolk sac. Does the yolk sac appear normal? Does the amount of yolk remaining in the control and experimental seem about equal?

  Step Cut off the yolk sac, dry your embryo lightly on paper toweling, and weigh it to the nearest ½ gram. Put the weight of your embryo in the appropriate column on the blackboard, and also record it.in your notes.

  Do your results so far suggest that:

a) thiourea prevents the manufacture of the thyroid hormone in the chick embryo, as it does in mature mammals? and that
b) an adequate supply of thyroid hormone is necessary for normal development?

Explain whether you think your results prove points (a) or (b). In what other ways could your results be interpreted?

Note: The remaining control and experimental eggs should be placed on separate shelves in the incubator, so that if any hatch, they can be distinguished from each other.

3rd and Subsequent Days Laboratory Procedure

Beginning 9 days after the thiourea injections (i.e., when the embryos are 20 days old). If any dry, fluffy chicks are found, remove them; be careful to keep control and thiourea-injected chicks separated. (Do not remove damp chicks the newly hatched chick must remain in the incubator at least 6 hours to dry before it can be safely removed). If any chicks have piped the shell but have not emerged, mark the shells for later identification and leave them in the incubator. The number of hatched chicks and newly piped eggs should be recorded. (***Note:*** Chicks may be safely left in the incubator overnight, if necessary).

The hatched, dry, chicks should be weighed and the weights recorded. The chicks should also be inspected carefully. Do the thiourea-injected specimens show any abnormalities? Are they as alert as the controls? Do they chirp as loudly? Can they stand and move as easily as the controls? Art there any defects in their feathering?

7th Day

Study the remaining chicks previously injected with thiourea, Review your notes on the 18-day thiourea-treated embryos that you examined 5 days ago. Review the later stages of the development of the chick embryo.

Laboratory Procedure

If some of the thiourea-injected chicks have not hatched, examine the eggs to see if there are any signs of hatching in particular, is the egg shell piped, or cracked? Now get out your preserved 18 or 19 day embryos from the study of normal development, and also your 18 day thiourea-treated embryo. Then open the egg by the technique used before for late stages. Note whether the beak has penetrated into the air space.

After freeing the embryo and its membranes from the shell, repeat the observations you made at the 18-day stage. Note especially whether tie yolk sac has been, or is being, retracted into the abdomen, since this is the first sign that hatching is about to take place. In your notebook, record the appearance of the embryo, weigh it, and note down the weight. Preserve the embryo, if desired:

If any thiourea-injected eggs still remain in the incubator, they may be left there to see if they will hatch later.

We can compare the thyroid glands of control and Thiourea treated chick embryos. Technical for removal of thyroid glands of chick embryo

To dissect out thyroids, lay the embryo on its back and slit open the skin of the neck from the breastbone forward, using fine scissors or a sharp-pointed forceps for this operation. Strip away the superficial muscles so that the windpipe is exposed. Near the base of the windpipe, you will see some large blood vessels running outward. The thyroid glands (two separate bodies in the chick) lie below (dorsal to) these blood vessels.

The identification of the thyroids is complicated a little by the fact that a chain of large nodes, which constitute the thymus gland, lie along both sides of the windpipe in this region. These thymic nodes are, however, pale, soft, spongy and irregular in shape. The thyroids, on the other hand, are pink, firm, have a regular, ovoid shape, and each is encased in a firm capsule.

To find one of the thyroid glands, push the musculature away from the side of the windpipe and firmly grasp the tissue surrounding the thyroid with your forceps. Pull out the gland, along with the tissue enclosing it; this tissue can be cleaned off after the gland has been removed from the body. Put the gland in the moist chamber, then repeat the operation on the other side of the body.

If an analytical balance or a fine torsion balance is available, weigh the pair of thyroids you have dissected out. Weigh them quickly to minimize drying. Speed is more important than accuracy in this case. If you do not have a balance, invert your petri dish, mark it with "C" (if yours was a control embryo), or "TU" (if injected with thiourea). Then look over the assembled thyroids. Is there any apparent size difference between the thyroids of the C and TU groups? Make a record of your observations.

CHAPTER 25

Basic Technique of Embryo Recovery Culture and Transfer in Mammals

Culturing mammalian embryos has been the most challenging problem. The reasons are obvious conditions of normal development which take place in oviduct and uterus are not easily duplicated. Recent innovations of in vitro fertilization and embryo transfer have been achieved only after extensive studies and by developing sophisticated culture methods.

Embryos may be collected from animals which have ovulated naturally or which have been induced to ovulate by administration of gonadotropins. This latter method allows greater flexibility in planning the experiments and usually results in an increased number of eggs ovulated. After recovery it is essential to place the eggs or embryos into an appropriate culture medium under controlled environmental conditions as quickly as possible.

Several investigators have developed their own chemically defined media for the culture of mouse pre-implantation embryos through fertilization and from the zygote to the blastocyst. All media are rather simple. They are based on a bicarbonate-buffered Krebs-Ringer type salt solution supplemented with nutrients (pyruvate, lactate, glucose), antibiotics and a high molecular weight component such as bovine serum albumin. More complex, serum supplemented media have been developed for the culture of blastocysts past the implantation period, for the culture of post-implantation stages, or for the establishment of embryo derived stem cell lines.

Table 25.1: Composition of media suitable for the culture of fertilized mouse eggs to blastocysts.'

Biggers et al.	Whittingham (1971)	Brinster (1971)	Ehitten (1972)	Hoppe (1971)	Mullen and (Personal Commun.)	Whitten (1971)
NaCl	554.0	553.4	554.6	514	640.0	515
KCl	35.6	35.6	35.6	36	. 35.6	36
$CaCl_2$	18.9	18.9	-	-	-	
KH_2PO_4	16.2	16.2	16.2	16	16.2	16
$MgSO_4$ 71120	29.4	29.4	29.4	29	29.4	29
$NaHCO_3$	210.6	210.6	210.6	190	190.0	211
Glucose	100.0	100.0	100.0	100	100.0	100
Na-lactate (DL)	241.62	260.8	280.0	2422	-	2422
Ca-lactate $5H_2O$	52.7	-	-	53	52.7	53
Na-pyruvate	2.8	3.8	5.6	3.5	2.5	4

Biggers et al.	Whittingham (1971)	Brinster (1971)	Ehitten (1972)	Hoppe (1971)	Mullen and (Personal Commun.)	Whitten (1971)
BSA	100.0	400	500	300	300	300

All ingredients are indicated as mg/100 ml. All media in addition contain K-penicillin-G (100 U/ml) and Streptomycin sulfate (50 mg/ml). The media by Whitten, Hoppe and Mullen & Whitten also contain phenol red, 0.1 ml/100 ml (Difco).

Eggs are cultured in an atmosphere of 5% CO_2, 5% 02, 90% N_2.

In vitro culture systems are widely used to study the direct effects of drugs on embryonic development independent of possible secondary effects exerted by the maternal environment. Moreover, the manifestation of developmental mutations or of induced genetic lesions can be observed throughout cleavage and implantation.

For completion of development to term, cultured pre-implantation embryos have to be transferred back into the oviducts or the uteri of a. recipient foster mother, usually a pseudopregnant female. Pseudopregnancy is induced by mating naturally ovulating' females with vasectomized males. Embryos are transferred via injection into the surgically exposed and punctured female reproductive tracts. The techniques of embryo transfer may be used to investigate the importance of the maternal environment, Embryos of a particular genotype may be transferred into a pseudopregnant foster mother of a different genotype, not containing any embryos, or alternatively they may be mixed with the embryos of a normally mated pregnant recipient female. The routine execution of embryo recovery and transfer as well as the availability of reliable culture conditions are prerequisites for successful experimentation with mouse preimplantation embryos. Each step should be trained and be firmly established in the laboratory before complex experiments are to be executed.

Materials

Tools and Equipment

The equipment for embryo recovery, culture and transfer includes: A stereo microscope equipped with transmitted illumination and magnifications ranging from about 10 X to 60 X (for embryo recovery and collection). A stereo microscope on a universal stand with incident illumination and magnifications ranging from about 5 X to 30X (for embryo transfer). A warming plate set at 37°C, equipped with a small plastic hood connected to a supply of a gas mixture composed either of 5% CO_2 in air, or of 5% CO_2, 5% 02, 90% N_2.

For microsurgical procedures in addition: An inverted microscope equipped with differential interference illuminations and objectives 10X and 20X. Two micromanipulators, one to the right, one to the left of the inverted microscope. Microscope and manipulators should be fixed onto a vibration-free table.

Moreover, the laboratory should be equipped with a CO_2 incubator if eggs or embryos have to be cultured for prolonged periods of time.

The small tools include: (a) Dissecting tools (scissors, forceps, needles) of appropriate size to operate mice, including a wound clip applicator and wound clips. (2) Embryological watch glasses or nontoxic plastic dishes. All glassware has to be thoroughly washed and rinsed free of detergents. (3) A mouth-controlled pipette. This is a universal tool and plays a central role in handling embryos. It consists of a mouth piece, a flexible tubing. An adapter and a glass capillary inserted into the adapter. Capillaries with an inner diameter of 100-150 mm are drawn by hand from 50 ml Pyrex Disposable Microsampling Pipette (a complete applicator mouth piece, tubing and adapter is supplied with each box). Drawn capillaries are sterilized before use and stored in closed containers. The mouth-controlled pipette serves to flush the oviducts or uteri, to pick up and transfer eggs or embryos throughout experimentation, including the transfer into the oviducts or the uteri of foster mothers.

Mouse Husbandry and Mating

Mice are kept in well ventilated rooms at constant temperature and with a normal day/night cycle. Depending on the particular experiment in mind, the mice may be chosen from a wealth of genetically defined inbred strains. If, however, genetic uniformity is not necessary, one of the many available outbred strains may be used. Alternatively, hybrid animals derived from crosses of inbred animals may be used, as they combine genetic uniformity with hybrid vigor. However, upon breeding such animals produce heterogeneous litters.

Mice reach full sexual maturity at an age of about 6-8 weeks, depending on the strain used. From this age, males and females may be used for matings. It is convenient to house the males singly in individual cages throughout their reproductive period to prevent fighting. Ovulation and mating are related to a diurnal cycle and usually take place around the midpoint of the dark period. Fertilization will occur shortly thereafter. Therefore, if fertilized eggs are needed the next day 2-3 randomly chosen females may be mated with each male overnight,, Successfully mated females can be detected the next morning by the presence of a usually prominent vaginal plug, formed by the solidified whitish secretions of the males vesicular and coagulating glands. Plugs can easily be detected by inspecting the vagina with a blunt spatula (see Fig. 25.1 & 25.2). To enhance the ratio of mated females, which is usually quite low on the first and the second nights, but quite high on the third night if males and females are left together, two methods may be applied: (1) The selection of females which are in proestrus and therefore ready for mating, and (2) the pretreatment of females with gonadotropic hormones.

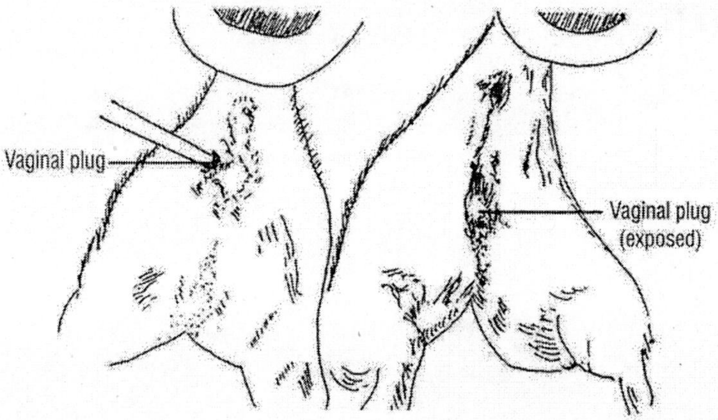

Fig. 25.1: Vaginal plug in contrast to the female to the right, the female to the left exhibits a prominent vaginal plug, indicating a successful mating the previous night.

Females have a normal cycle of about 4-5 days. The appearance of the external vagina changes drastically during the cycle, from pale and dry in metoestrus to reddish-pink and moist in proestrus. With some experience females in proestrus can he selected from the stock late in the afternoon. Selected females should be housed singly with males, as most of them will be mated. Preselection is important when females are to be used as foster, mothers.

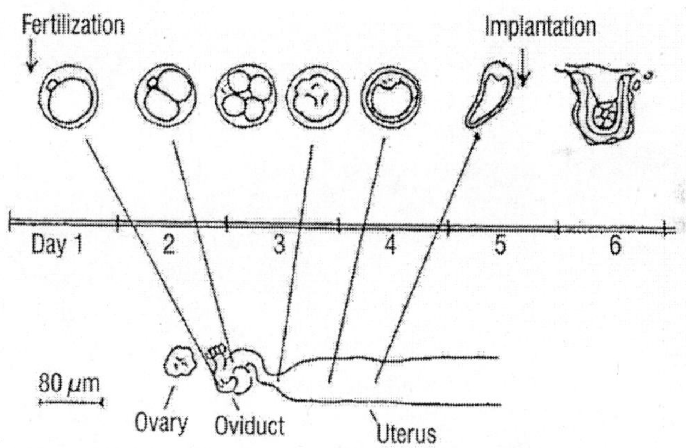

Fig. 25.2: Early development of the mouse embryo with regard to its location in the female genital tract.

Alternatively, immature and mature females may be pretreated with hormones in order to stimulate egg maturation and ovulation. An injection of 5-10 IU of PMS (pregnant mares serum) is followed by an injection of 5 IU HCG (human chorionic gonadotropin) 46-48 h later. Both, hormones are dissolved in saline at concentrations of 50 IU/ml. Injections are given intraperitoneally.

Reliable hormone preparations are obtained from Intervet (PMS = Chorulon, HCG = Folligon). If HCG is given around noon, ovulation will be induced at around midnight (about.12 h after-HCG injection), thus at about the time when spontaneous ovulation occurs. Immediately after injection of HCG the females are paired one by one with a male. Again, successful mating is indicated by the presence of a vaginal plug.

The day of detection of the plug is called day 1 of gestation and accordingly, with regard to the embryo, day 1 of development.

Procedure of Preparation of Media

The composition of several media suited for the culture of mouse zygotes to the blastocyst stage are listed in the table I. In preparing media for the culture of mouse embryos it is essential to use water of the highest purity. This can either be double or triple distilled water or, water demineralized in a combined activated-charcoal/mixed-bed ion exchange system. Media should be prepared in thoroughly cleaned and rinsed glass or nontoxic plastic bottles. The ingredients should be reagent grade or of the highest purity available. Prolonged storage of media should be avoided, therefore we prepare fresh medium (modified SECM) once a week. The dry ingredients with the exception of bovine serum albumin (BSA) are weighed in a clean bottle. One hundred milliliters of water are added and the salts dissolved by gentle shaking before adding phenol red (0.1 ml, Difco). The medium is then gassed for about 10 min with 5% CO_2 in air by bubbling gas mixture from the tip of a disposable pipet into the salt solution. Twenty milliliters of the gassed salt solution are placed into a Nalgene filter unit (0.45 mm pore size) and the membrane is allowed to moisten. The solution is forced through the filter-by positive pressure from the supply of gas mixture and discarded. Two hundred and forty milligrams of bovine serum albumin and dissolved in the remaining 80 ml of salt solution (= 3 mg/ml) to make up the final culture medium. The medium is filtered and stored in gastight bottles filled with the previously used gas mixture at 4°C. Egg culture media may contain 0.1 mM EDTA (3.7 mg/100 ml), as this has been shown to be beneficial for the development of mouse 1-cell embryos. To prevent pH changes upon prolonged exposure to the air, media used during manipulation may be modified by, substitution of 25 mM Hepes buffer (pH 7.4) for the bicarbonate. To render eggs more resistant-to mechanical damage, the medium used during the micromanipulation period may in addition be supplemented with cytochalasin B (5 mg/ml). Cytochalasin B suppresses cytokinesis without affecting the progression of the nuclear cycle.

Therefore, if eggs have to develop normally they have to be washed and incubated in normal medium. If however, tetraploidy or polyploidy of the embryonic genome shall be induced, fertilized eggs or 2-cell stages may be incubated for one or more rounds of DNA replication before they are further incubated in normal culture medium.

Recovery of Eggs and Embryos

The early development of the mouse embryo with regard to its location within the genital tract of the female is schematically represented in figure r. From there it follows that eggs (on day 1 of development) are recovered from the ampulla of the oviduct, 8-cell embryos (on day 3) from the uterotubal junction, blastocysts (on day 4) from the uterus. Accordingly, either the two oviducts, the oviducts with a small piece of the upper uterus each, or the two uterine horns have to be excised from the abdominal cavity of the female. For this purpose the animals are killed either by cervical dislocation or by quick suffocation in CO_2. A closed box, divided in two compartments by a grid containing some dry ice in one compartment, is a suitable device. The dead females are opened from the ventral side by pulling apart the ventral skin and' by opening the abdominal wall with scissors. The exposed small intestine is pushed towards the thoracic cage. The dorsally located right and left uterine horn can easily be seen.

Oviducts are removed by pulling the uterus in order to stretch the ovary and the oviduct. With small scissors the expanded mesentery between ovary and oviduct is cut as illustrated in Fig. 25.3. There is no danger to injure the tube because the ostium (the open ovarian end of the oviduct) simply reaches into the ovarian bursa with no direct connection to the ovary. The oviduct is carefully lifted and a second incision is made at the uterotubal junction. The free oviduct is placed in a dish containing some medium. If 8-cell embryos have to be collected a small piece of the uterus should be removed with the oviduct. If blastocysts have to be recovered the uterine horns are freed from all mesenterial fat and dissected by two incisions, one near the tubal end the other next to the vagina.

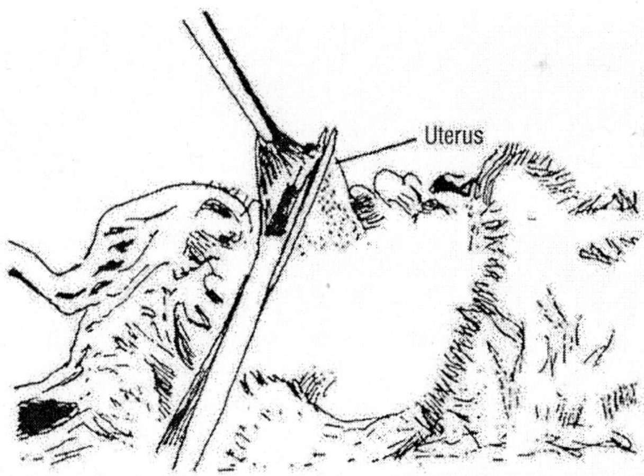

Fig. 25.3: Excision of an oviduct: The uterus is pulled out of the open body cavity in order to stretch the oviduct and the ovary. A first incision is made in between the oviduct and the ovary. The oviduct is cut free from the uterus by a second incision across the uterotubal junction.

If freshly fertilized eggs have to he recovered to tubal end, then it is necessary to flush the oviduct. Especially during the morning of day 1 the eggs remain jointly within the transparent swollen ampulla where they can easily be detected upon inspection of the oviduct. In that case the ampulla simply is disrupted with a sharp needle. Due to the decompression of the ampulla the eggs float out into the medium. At this stage the eggs usually are still surrounded by follicular cells. These have to be dispersed by adding a few crystals of lyophilized bovine hyaluronidase to the medium. Fertilized eggs are easily recognized and distinguished from unfertilized eggs by the presence of a prominent second polar body.

If the eggs have already left the ampullar region (usually in the afternoon of day 1) or if 2 to 4 cell embryos have to be recovered, the oviduct has to be flushed. This is done by using a mouth-controlled capillary prefilled with medium. The tip of the capillary is inserted into the ostium. The outer diameter of the capillary should tightly fit the inner diameter of the oviductal tube. By enhancing the air pressure in the mouth-controlled pipette medium is forced through the coiled tubing of the oviduct and the embryos flushed into the dish.

Blastocysts are recovered by inserting the tip of a somewhat larger pipette into the lumen of the excised uterus. The end of the uterus is squeezed around the pipette with a pair of forceps and medium is forced through the uterus as described for oviduct.

Immediately after flushing the eggs or embryos are collected using a mouth-controlled capillary and transferred in a small amount of medium into fresh medium to get rid of the debris flushed from the reproductive tracts. Washing is repeated two to three times, especially when hyaluronidase had been added to the medium to remove follicular cells. Finally, the eggs or embryos are incubated in drops of medium covered with paraffin oil for culture, microsurgery or transfer.

Egg and Embryo Culture

For culture, mouse eggs and embryos are incubated in one of the media described, kept as 37°C in an atmosphere of 5% CO_2 in air, or occasionally 5% CO_2, 5% 02, 90% N_2. Two types are convenient and give satisfactory results: (1) Glass culture tubes (12 x 75 mm containing 1 ml of medium, gassed with gas mixture and filled with a rubber stopper. Suet tubes are autonomous with regard to the surrounding atmosphere. They are appropriate in till absence of a CO_2 incubator or when a complicated gas mixture has to be used. However, such tubes are less convenient with regard to the observation and recovery of cultured embryos. (2 A most convenient culture system are drops of medium covered with nontoxic paraffin o silicon oil. Microdrops of medium are placed directly onto the bottom of a plastic tissue culture dish by using a medium filled syringe equipped with a fine, needle. Subsequently, the dish i carefully filled with oil until the drops are completely covered. The microdrops remain stable in place. Eggs or embryos are placed singly or in small groups into the drops (or remove from them) with a mouth-controlled capillary. The fishes are incubated in an appropriable atmosphere for pre-equilibration and culture. The paraffin or mineral oil overlaying the medium prevents evaporation, rapid gas exchange and contamination of the medium by airborne germs. Before use the oil is equilibrated by mixing and stirring 5 parts of oil with 1 part of medium without BSA for 24 h [Hoppe and Pitts, 19731. After the separation of the two phases the medium is discarded. Equilibrated oil may be stored and used for several weeks. Two types of oil have been tested in our laboratory with equally satisfactory results paraffin oil and silicone oil.

Surgery, Narcosis

Surgical operations include egg and embryo transfer into the reproductive tracts of appropriately timed foster mothers as well as vasectomy of males. During surgery the mice have to be anesthetized for a few minutes. This is conveniently done by an i.p. injection of either a solution of sodium pentobarbital at a concentration of 0.6 mg/10 g body weight, or of a tribromethanol solution at a concentration of 3.5 mg/10 g body weight. After surgery, the mice are placed under a hot lamp or on a warming plate to recover from the drop of the body temperature during narcosis. Mice are rather resistant to infection. Nevertheless, instruments and skin should he washed with 70% ethanol before operation. Once again fully awake, the animals are housed singly in autoclaved cages for further recovery.

Egg Transfer

Fertilized eggs (or 2-cell stages) are transferred into the ampulla of a pseudopregnant foster mother on day 1 of pregnancy. For this purpose a patch of hair is removed by shaving the back of the anesthetized female. The skin is cut open along the mid-dorsal line for about 1 cm. The opening may be retracted to either side to reach both oviducts in turn. The fat pad covering each ovary is usually readily visible across the thin lateral abdominal wall, which is incised above the fat pad for about 5 mm. The fat is seized and pulled to the Outside of the body. Here it is immobilized with a pair of forceps, otherwise it will retract into the abdominal cavity. The female is placed under the dissecting microscope with the Universal stand. Under the other microscope, the embryos are picked up into the tip of a fine, mouth-controlled capillary kept with the right hand. When moving to the prepared female the right hand in addition seizes a pair of sharply pointed watch marked forceps or a very fine needle, adequate to puncture the oviduct. Under the dissecting microscope the oviduct is inspected and the ampulla region exposed. From the left the oviduct is firmly grasped with fine forceps between the ostium (which is inside the ovarian bursa) and the ampulla. The fine forceps are slightly turned in order to bend the oviduct is carefully punctured, the forceps are dropped, and the tip of the capillary containing the eggs is inserted into the fine puncture hole (Fig. 25.4). The eggs are blown into the oviduct in a small volume of medium. After transfer the fat pad and with it the ovary, the oviduct and the uterus are placed back into the abdominal cavity and the operation repeated on the other side. When the operation is finished, the skin wound is closed with a clip. The incisions in the abdominal wall are left open.

Fig. 25.4: Egg transfer into the ampulla of an oviduct: The ovary and oviduct are exposed across a dorsal incision. The oviduct is punctured and the eggs transferred into the ampulla by means of a fine capillary.

Morulae/ Blastoeyst Transfer

Morulae or blastocysts are transferred into the uteri of females late on day 3 of pregnancy. The operation is executed as described for the egg transfer, with the exception that the embryos are deposited in the uterus near the uterotubal junction. Care has to be taken not to irritate the uterus by extensive squeezing or puncturing with a large needle.

Upon vasectomy, portions of both vas deferens of a male and removed surgically, as illustrated in figure 8. The anesthetized Male, ideally of an age of 6-8 weeks, is positioned ventral side up in front of the investigator. The scrotal sac is thoroughly disinfected with 70% ethanol and opened by a single inidline incision. Care. Has to be taken not to hurt the anus. One by one the usually retracted testicules are pushed from the ventral side of the abdominal cavity into the scrotal sac. The contents of the scrotal sac are firmly squeezed with blunt forceps between the tail of the epididymis and the testicule. Across the transparent abdominal wall the vas deferens can be spotted on the outside of the testis as a straight whitish tubule accompanied by a fine blood vessel. With a pair of watchmaker forceps the abdominal wall is punctured, the tubule grasped and a loop is pulled across, the puncture hole (Fig, 25.5 & 25.6). The vas may be ligated once using sterile suture. This step may however be omitted. The vas is cut distal from the ligation and a 1 to 2 cm long portion is pulled out of the abdomen and excised. The wound is closed with a single suture.' Males should be given a week to recover before they are used for mating. Vasectomized males exhibit normal mating performance. They may be used for up to 1 year after operation.

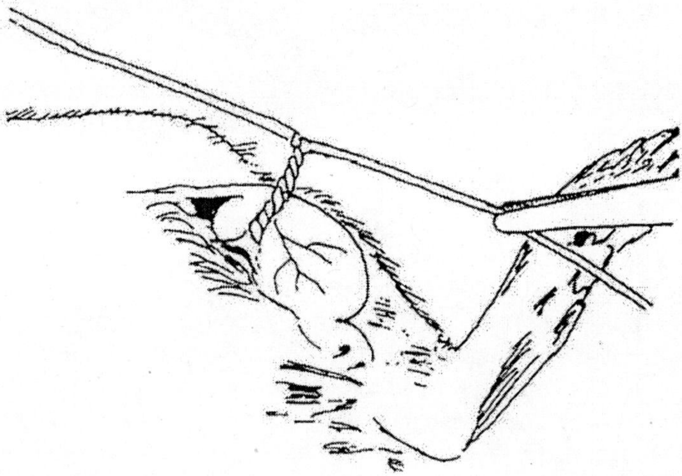

Fig. 25.5: Vasectomy: the vasdefrens is exposed across a puncture hole in the scrotal abdominal wall and ligated.

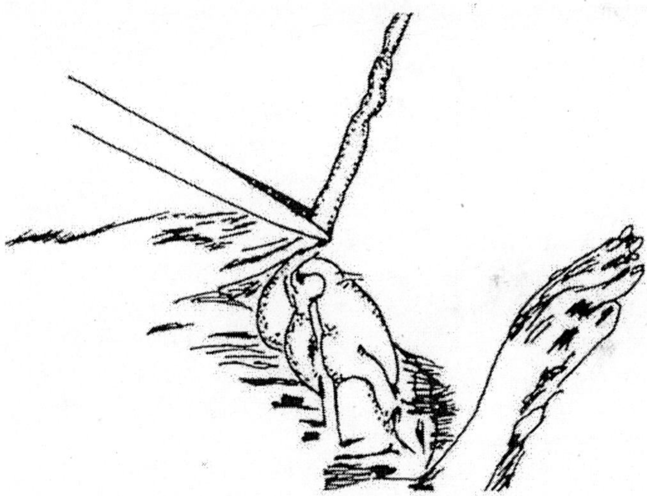

Fig. 25.6: Vasectomy: A major part of the vasdefrens is excised.

CHAPTER 26

Early Embryology of the Mouse

Thy study of mammalian embryology is hampered by two factors, first the availability of the material and second, the usual large size of the embryos that have been used. The mouse meets both of these complications by being readily available for any laboratory and being relatively small throughout development.

Any species or strain of mouse may be used. The description here will be based upon the pure strain of white albino mouse which is widely used for research. It is not a hybrid and thereby avoids the wider variations in development attendant upon great heterozygosity. The stages described are essentially the same for any strain, the difference being in rate of development as it is presumed that the hybrid may grow slightly more rapidly than the pure strain. The form described here has a 21 day gestation period, as does the rat, so that the development of these divergent strains may be compared.

Securing of the Material

The mouse has a 5 or 5-½ day estrous cycle, and will accept the male only at the height of estrus, which generally occurs between 10 P.M. and 1 A.M. The most accurate method of timing conception is to observe the mating, but short of this it is possible to estimate the conception time rather closely, within several hours. Except for the early fertilization stages, such accuracy is not necessary.

Generally a box of 5 or 6 females is provided with a sexually mature male at 5 P.M. and the next morning at 9 A.M. each female is examined for the presence of a vaginal plug, a certain sign of successful copulation. The plug consists of sections from the vesicular and coagulating glands of the male. The vesicular glands may be known as the seminal vesicles. They do not contain spermatozoa hut do produce a secretion. The coagulating glands are the prostates whose secretion, when mixed with that of the vesicular glands, forms the coagulum that blocks the vaginal Orifice. A vaginal smear, made by inserting a flamed pipette (filled with 0.5 cc. of physiological saline) into the vagina and sucking out sonic of the contents and examining with dark field illumination under the microscope for living spermatozoa, is more tedious but another method of selecting those females that have mated.

Since mating could occur between 5 P.M. and 9 A.M., but is more likely to occur between 10 P.M. and 1 A.M., there is some variation in the early stages. This could be as much as 16 hours. The first cleavage occurs after 24 hours of fertilization, so that one can easily obtain the pre-cleavage stages of fertilization.

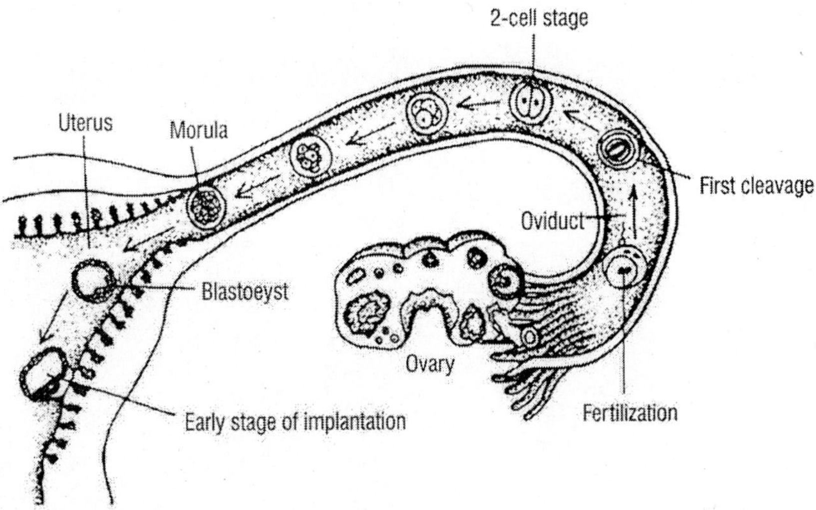

Fig. 26.1a: Development of a mouse embryo from fertilization to implantation.

Fertilization

Fertilization occurs in the upper end of the oviduct, closely approximating the ovary with fertilization, the zona pellucida swells. Polyspermy is not known. Perivitelline space develops between the zona pellucida and vitelline membrane, sperm head enters the vitellus. Within 2-3 hours the second polar body is formed and everted into the perivitelline space and toe pro-nuclei (male and female) begin to migrate toward each other (Fig. 2.1.1a)

Cleavage

The first two cleavages of the mouse „egg occur while it is still within the oviduct, and generally one will find groups of 5 or 6 such cleavage stages close together in each oviduct. The first cleavage occurs at about 24 hours after fertilization, but since this may vary in time one estimates it to occur at about midnight so that by 9 A.M. development is considered to be already ½ day. Thus, on successive moorings, one estimates the ages as 1½, 2½, days etc.

Cleavage itself takes only about 10 minutes so that mitotic figures are rarely seen. The inter mitotic time is many hours during the early cleavages. Each successive cleavage occurs more quickly and by 2½ days after fertilization the embryo is a 16 cell morula (solid ball of 16 similar cells). Cleavage is not synchronous so that intermediate numbers of cells may be found, between 4 and 16, for instance by 3+ days the eggs descend from the oviduct into the uterus, but may still be clustered. One of the unexplained situations then follows, namely the rather even distribution of the embryos for the length of the uterus, spacing so that each embryo has adequate implantation and growing room. Never does one find two closely implanted embryos. (Fig. 26.1b)

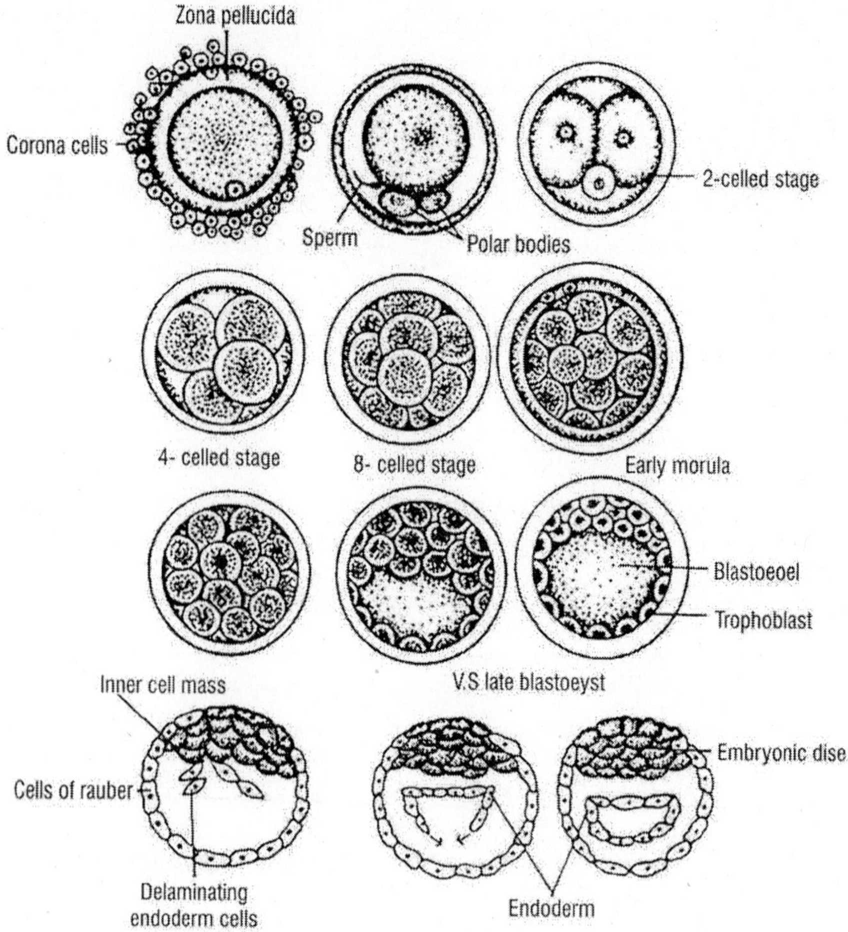

Fig. 26.1b: Gastrulation in mouse. Formation of endoderm.

The Blastula Stage

Shortly after reaching the 32 cell stage the morula begins to acquire an eccentric cavity, filled with a fluid, known as the blastocoel (segmentation cavity). This enlarges rather rapidly while the cells at one side appear to multiply and cluster to form the inner cell mass. The related single layer of cells which continue around the blastocoel will become the trophectoderm. While these, changes are occurring the embryos descend into the uterus and take up positions for implantations equidistant from each other. (Fig. 26.1b)

Implantation of the Embryo

From the innermost epithelial lining of the uterus outward are the relatively thick mucosa, the inner circular and outer longitudinal muscles, and finally the mesometrium which attaches the uterus to the dorsal body wall. Both the circular and longitudinal, muscles function to give the uterus undulating movements, which undoubtedly help in moving the embryo along through the lumen. Each blastula becomes located in an epithelial crypt, quite equidistant from adjacent blastulae. The nuclei in the vicinity of the uterine epithelium of the blastula then begin to disintegrate and within 15-20 hours this epithelium is entirely sloughed off. This will be the implantation site for the particular blastula examined. The trophectoderm cells of the blastula make the contact with the epithelium, and there begins a response in the nearby mucosa which results in sufficient swelling by 5 days to be noted with the naked eye. This swollen mucosa is known as the decidua; which will be an almost immediate source of nutriment for the growing embryo.

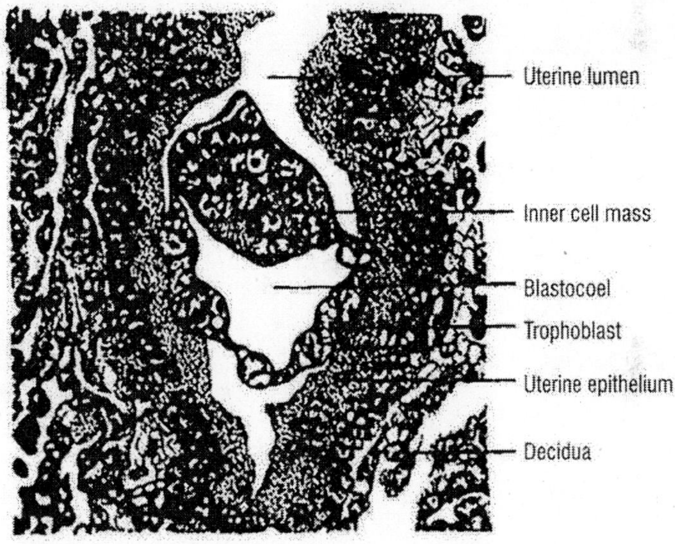

Fig. 26.2a: Section of mouse blastocyst lying in a crypt and just about to embed.

The zona pellucida disappears from the egg during the fourth or fifth cleavages, even though there appears to be no actual increases with blastulation, and as soon as implantation is achieved, the embryo appears to accelerate its growth processes. The blastocoelic fluid distends the blastula so that occasionally one can find a blastula making contact on two sides of the uterine epithelial crypt. At the same time the inner cell mass, cells on one side of the blastula, increase in number giving it a multi-layered appearance as compared with the trophectoderm. The blastocoel is now known as the yolk cavity and the inner cell mass as the egg cylinder and the embryo as a blastocyst. It is the trophectoderm which makes the invasion into the uterine mucosa, establishing nutritional contact with the maternal tissues. (Fig. 26.2 a&b)

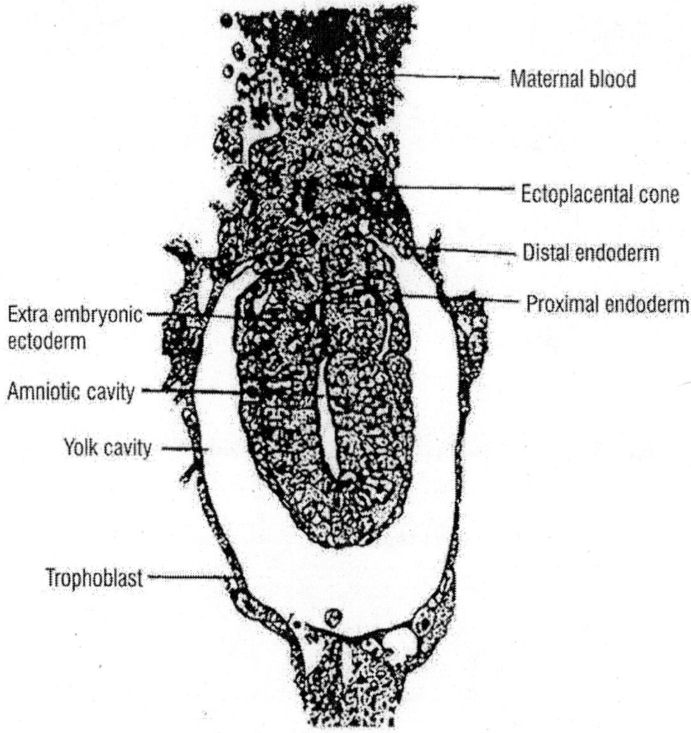

Fig. 26.2b: Section of a mouse embryo at a slightly older stage than in Fig 26.2a. The egg cylinder has lengthened.

The Primary Germ Layers

The inner cell mass almost immediately develops two types of cells, the thinner and more darkly staining layer next to the yolk cavity will be the endoderm, and all other cells of the blastocyst are early embryonic ectoderm. The trophectoderm (troph = Greek for nourishment) does not give rise to any embryonic structures. The embryo arises from the inner cell mass.

The endoderm of the inner cell mass, at about 4½ to 5 days, gives rise by proliferation to a column of cells which appear to come from the junction of the endoderm and the trophectoderm and project into the yolk cavity. These cells will eventually line the yolk cavity, closely applied to the outer trophectoderm, and will be known as the distal endoderm to distinguish them from the proximal endoderm of the inner cell mass. During this process the embryonic ectoderm of the inner cell mass has thickened, and the outer-most cells appear to form another layer known as the extra-embryonic ectoderm. These cells appear to invade the maternal mucosa and destroy its blood vessels, forming the ectoplacental cone. The embryonic ectoderm has, in the meantime, developed its own cavity known as the pro-amniotic cavity. (Fig. 26.2b)

The embryo at 6 days consists of four rings of cells. The innermost is the embryonic ectoderm, the next the proximal endoderm, separated by the yolk cavity from the distal endoderm and the rapidly disappearing trophectoderm. With this disappearance one finds many invading giant cells. While the trophectoderm is extra-embryonic, it disappears and the term is reserved for that ectoderm giving rise to the extra-placental cone and collectively known as the egg cylinder. The extra embryonic ectoderm not comprising the extra-placental cone then acquires an elongated shape. The cells of the ectoplacental cone will give rise to part of the placenta. It appears to be actively invasive.

At this point it must be emphasized that the relative positions of the primary ecto and endoderm are the reverse of what one would expect from the study of other vertebrates (except the rat, rabbit, and guinea pig). The embryonic ectoderm is the innermost layer of cells, and by 6½ days the mesenchyme (embryonic mesoderm) will begin to appear intermediate between this inner ectoderm and the outer endoderm. After all, the confusion (if any) arises from human prejudice in the naming of presumptive layers, while the mouse allocates its cells and layers with dependable regularity and always comes out "a mouse".

As in most forms, the embryonic mesoderm arises by proliferation from prior existing embryonic ectoderm, near the junction with the extra embryonic ectoderm. This is the site of the primitive streak. At first the mesenchyme consists of loosely spaced cells which work their way between the ecto and endoderm, particularly in the vicinity of the junction of embryonic and the extra embryonic ectoderm (the future anterior of the embryo). Some mesenchyme also invades the region of the extra embryonic ectoderm later to form the double layered extra embryonic yolk sac, which will entirely envelop the embryo.

Topographical Considerations

The original site of the mesenchyme identifies the position of the thickened ectoderm known as the primitive streak. This also is found toward the posterior of the egg cylinder, but obviously to one side. Since the original position of the inner cell mass is later marked by the ectoplacental cone, the embryo develops in a direction opposite to that of the inner cell mass, but within the yolk cavity. The embryo always develops with its anterior-posterior axis perpendicularly to the mesometrium, and the dorsal-ventral axis parallel to the mesometrium. Thus, if one cuts the gravid uterus lengthwise he is apt to cut the early embryo either transversely or frontally, and a transverse cut of the gravid uterus should provide a sagittal section of the embryo. One must remember, however, that by the time the embryo is shaped like a capital "S", transverse cuts may be misleading. Eventually, by an inversion of the fore-and hind-gut, -the embryo assumes the capital “C” shape so characteristic of the higher vertebrates. (Fig. 26.3a, b, c&d)

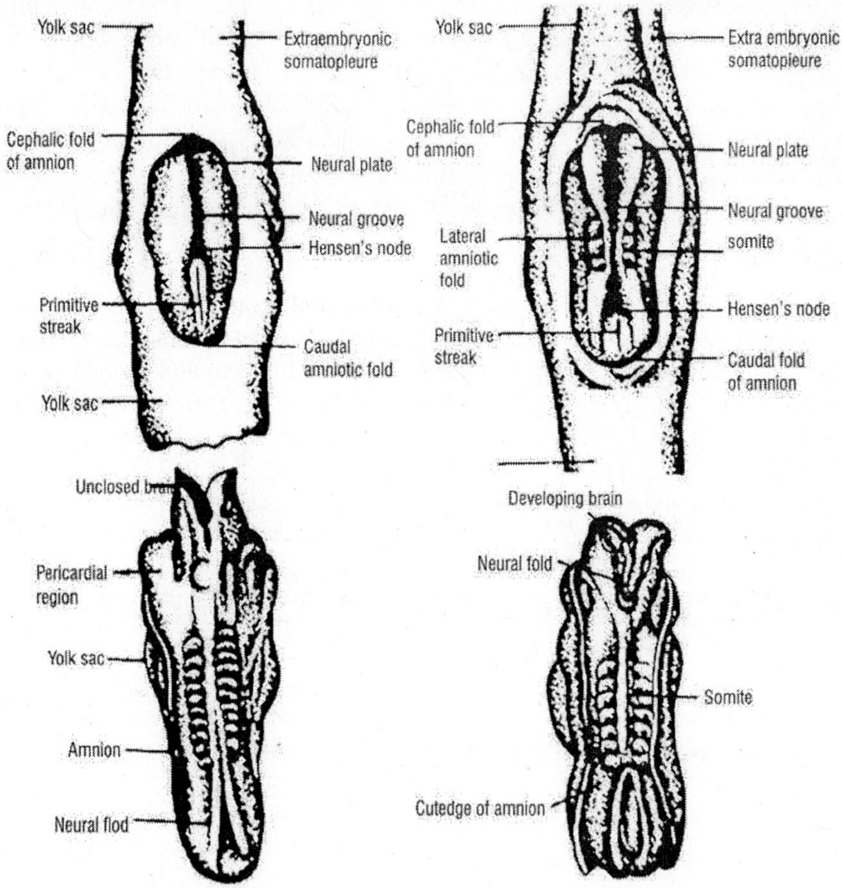

Fig. 26.3: (a) The embryo is showing the first appearance of neural groove. (b) Embryo at the time of appearance of the first somites. (c) 7 somite stage embryo. (d) 10 somite stage embryo.

Membranes and Cavities

The pro-amniotic cavity has already appeared. It is surrounded by embryonic ectoderm and is internal. The mesoderm, appearing first between the ecto and endoderm at the junction of the embryonic and extra-embryonic regions, causes a bulging of cells into the pro-amniotic cavity. As this bulge acquires a cavity the combined meso and ectoderm form the posterior amniotic fold. In sagittal sections one cannot see it, but this mesoderm is continuous around the egg cylinder to form the lateral and finally the anterior amniotic folds. This mesodermal sheet, as it thickens, tends to constrict the egg cylinder at about its midpoint. The cavity or cavities which develop, largely in the posterior mesoderm, coalesce to form the exocoelom (extra-embryonic coelom), lined with embryonic structures are found toward one pole of the egg cylinder (ventral) and the extra-embryonic structures at the other (dorsal).

The original pro-amniotic cavity is now subdivided into the ventral amniotic cavity (ectodermal), the exocoelom (mesodermal), and the ectoplacental cavity (ectodermal). This latter cavity will ultimately disappear. The double (ecto and mesodermal) membrane which separates the amniotic cavity from the exocoelom is the amnion, and the similarly constructed chorion separates the exocoelom from the extra placental cavity.

The neural Axis

The head process develops at 7 days as an anterior prolongation of the cells from the dorsal primitive streak, probably largely ectoderm although originally in the position where one would expect mesoderm, between the ecto and endoderm. It is continuous around the ventral side-of the egg cylinder as squamous endoderm. This gradually becomes columnar endoderm as one views the sheet of cells around toward the ventral side of the embryo. The primitive streak then extends from the most anterior tip of the egg cylinder to the level of the amnion, or about 2/5. of the length of the cylinder. In transverse sections through the head process one can see that its cells are loose and appear to come from the inner, thicker ectoderm. They extend about 3/4 of the way around the embryo, but are much thicker at the dorsal side. They are contiguous with mesenchyme and can be distinguished only by virtue of their position and destination. It is noted, however, that the cells of the head process remain in contact with the outer endoderm while the mesenchyme adheres to the inner ectoderm. By 7.5 days the mesoderm has almost completely separated the ecto from the endoderm, including all but the mid-dorsal region of the embryo where the notochord axis will form. (Fig. 26.3a, b, c&d)

The notochord arises from the outer, thin endoderm first by a slight thickening and then an invagination to form a groove. The inverted groove of endoderm is then the primitive archenteron, with the cells pinching off above it to form the notochord. At this level there are no mesenchyme cells between the ecto and endodermal layers.

The neural groove appears directly above the region of the notochord, as an, inverted depression or trough in the thick ectoderm which extends forward in the mid-sagittal line as far as there is embryonic ectoderm. By 7.5 days it is easily identifiable as a deepened (inverted) groove with thickened sides, and by 8.5 days it is closed as a neural tube continuous with the embryonic brain.

The notochord develops during these neural changes, with its originally oval nuclei becoming re-oriented in a plane perpendicular to the ectoderm of the neural groove. The notochord retains its connections with the archenteron (gut endoderm) for a considerable time, but eventually (as in Amphioxus) the notochord becomes pinched off from the archenteron and forms the axis around which the embryo develops its central nervous and skeletal structures. Thus, the head process gives rise to the gut endoderm as well as the notochord.

The allantois of the mouse is not associated with the endoderm, as into many vertebrates, but is purely mesodermal, arising in association with the original exocoelom into which it grows, prior to acquiring, its cavity. By 7.5 days it appears as a globular mass apparently destined to fill the exocoelom. Its function, as in all vertebrates, is nutritional and respiratory. The chorion, composed of ectoderm and mesoderm, is thickened, tends to compress and all but obliterate the ectoplacental cavity. By 8 days the allantois becomes adherent to the chorion, thereby establishing continuity between the embryo (primitive streak) and the region of maternal contact (ectoplacental cone). Thus, the ground work is laid for close circulatory, nutritional and respiratory connections with the maternal tissues.

The foregut primordium appears first as a notch in the thick columnar endoderm toward the anterior limits of the egg cylinder, at about 7 days of development. The notch enlarges into the direction of the amniotic cavity, taking with it the overlying ectoderm, thus, forming a distinctive fore-gut pouch. The greatly thickened ectoderm anterior to this foregut is the head fold ectoderm. During the next day the foregut enlarges rapidly to form a deep pocket and then a pouch, which becomes constricted laterally and ventrally by the appearance of mesoderm, with the anterior intestinal portal being continually shifted posteriorly (caudally). Eventually the mesoderm will come to encircle the foregut as it is formed. Up to this stage the germ layers have appeared to be in a position opposite that of other vertebrates studies, so that this inversion of the endoderm brings it, and the surrounding mesoderm and ectoderm into the relationship one expects of the adult.

At the posterior margin of the primitive streak, beginning shortly after the invagination of ta, foregut, will appear the invagination of the hind-gut. It will form in the same manner as the foregut, developing a posterior intestinal portal and eventually it will become continuous with the foregut at the level of the last-to-form mid-gut. The stomodeum (invagination of head ectoderm) and proctodeum (invagination of caudal ectoderm) will eventually rupture through to the gut to form the mouth and the anus, first appearing at about 8 days. One may even see an oral and anal plate, points of contact of ecto and endoderm prior to rupture.

The head fold is accentuated by the developing foregut, as the ectoderm thickens and is pushed forward into the amniotic cavity. With neurogenesis this head fold and process then become the aggressively active differentiating center of the embryo.

By 7.5 days of development the embryonic mesenchyme has invaded almost all parts of the embryo and extra-embryonic structures, between the ecto and endoderm, and also constitutes the lining of the exocoelom. The allantois, invading the exocoelom, is entirely mesodermal. About the only region the mesenchyme cannot invade is where the endodermal notochord is being developed, in the mid-sagittal plane. Beneath the head fold and process the mesodermal sheets come together to form the primitive rudiments of the heart, anterior to the foregut as it develops. Later the heart is ventral to the foregut, due to the forward development of the head process.

On each side of the notochord the sheets of mesenchyme, which are continuous with the extra-embryonic mesoderm, become organized so that by 8+ days there appear clusters of mesoderm just lateral to the notochord, the primordia of the somites. While somite mesenchyme is at first contiguous with the more lateral mesoderm, it becomes partially separated by a constriction which demarks the lateral plate mesoderm that acquire, a cavity known as the caelorn. The first definitive somite appears at about 7.5 days and by 8.0 days there may have developed as many as 6 or 7 pairs. The first to develop is the most anterior and it appears to be actually further anterior than the primitive streak, dorsal to the developing foregut. The second appears next and just posterior to the first and becomes constricted off from it to form the characteristic somite block of cells seen in all developing vertebrates. Eventually there develop about 65 pairs of somites. It is likely that the first one or several lose their identity with the development of the head structures. During this development one can study the various stages in somite formation simply by examining the mesenchyme lateral to the notochord from the most anterior (most developed) to the most posterior (least developed) for several days of gestation from 8 days onward.

As in all typical vertebrates the primitive streak is the organizing center, forward from which develops the embryo. It gives rise, actually, to all three germ layers and is in this respect comparable to the dorsal lip of the blastopore of other forms. The allantois, usually arising from the floor of the hindgut, here arises from the posterior extremity of the primitive streak, still close to the hindgut. The embryo arises anteriorly to the streak and at its expense, but as the embryo develops it too, shows cephalic activity which provides the cephalic or head fold. But differentiation begins at the most anterior extremity of the embryo so that the more posterior primitive streak remains the most retarded or undifferentiated region even though it gives rise to the germ layer prerequisites of the embryo.

The coelom is developed as in other vertebrates, by a split in the lateral Mesoderm to form the intervening cavity. The amnion is a continuation of the somatopleure (ecto-and mesoderm). The splanchnopleure (meso and endoderm) continues as the membrane forming the floor of the exocoelom laterally. Ventral to the forming foregut the mesodermal cavity (coelom) becomes the pericardial cavity. The relationship of these precardial mesodermal parts can be seen best at about 8 days of development in serial sections.

Reichert's membrane has no counterpart in other forms, is peculiar to rodent development, is non-cellular, elastic, tough, and lies between the endoderm of the developing yolk cavity and, the outer trophectoderm. It may arise from the endoderm, and is the outermost, all-inclusive membrane of the embryo. But it is non-living. The amnion is structurally and functionally similar to that of other vertebrates, but the chorion has lost its protective function to the

Reichert's membrane and is insignificant. As the embryo becomes involved with maternal tissues this Reichert's membrane becomes obscured and probably disappears, being non-living. It may best be seen on day 6.5. By 8 days the amnion extends over the entire dorsal surface of the developing embryo, and later envelops the embryo so that it is free to float and develop within.

Since nutrition for the mammalian embryo is derived through the maternal tissues, there is no need for a yolk sac but in rodents it retains some vestige of its ancestral function by establishing a nutritional relationship between mother and embryo through at least a good portion of development. It is said the yolk cavity arises from the original blastocoel, but comes to lie between Reichert's membrane and the egg-cylinder, and that its membrane is that middle potion which comes from the extra embryonic splanchnopleure. As the embryo grows and its membranes and cavities expand, this yolk sac, by 8.5 days, envelops the very large exocoelom and all but the eventual mid-gut region of the embryo. When the midgut finally closes, the entire mass is enclosed by the yolk sac (amnion, embryo and all). The inner mesodermal layer of the yolk sac acquires blood islands at about 7.5-8.0 days, the precursors of the embryonic and extra-embryonic blood vascular systems, which are continuous.

The presence of a parasite; in the uterine cavity, in the form of a developing embryo, cause responses on the part of the uterus, which can be observed with the naked eye in the form 01 swellings as soon as 5.5 days after fertilization. Implantation begins at about 4.5 days so tha this reaction appears quite soon thereafter. This is known as the decidual swelling because it i; due entirely to changes in the uterus itself.

The 4.5 day embryos are suspended within the uterine cavity and for reasons not understood they come to be spaced equidistant throughout this lumen. Each embryo then is associated with the uterine epithelium, becoming attached often by the trophectoderm. A crypt develops around the embryo, which soon envelops it, cutting it off from the uterine lumen at 7.5 days. Whit there are some variations, generally the uterine tissue which finally closes the embryo off from the uterine cavity will give rise to much of the placenta, being located dorsal to the ectoplacental cone in most cases. By 8.0 days the embryo is completely cut off from the uterine cavity and for a while it obstructs the lumen completely at this level, the only remaining lumen being between implantation sites. By 9.0 days the lumen on each side of the embryo has reestablished connections, but ventral to the embryo, thus, representing a new special connection. Thus, the old uterine lumen is obliterated by the decidua basalis and eventual placenta, while the embryo will come to lie freely within a new uterine lumen, locate ventrally to the embryo outside of the decidua capsularis.

Changes in the, define mucosa are first seen at 5.5 days when giant cells appear ventral to the embryo in association with degenerating uterine epithelium.

At the opposite pole and probably derived from the ectoplacental cone there will then appear rather numerous giant cells outside of ire trophectoderm, characterized by extending protoplasmic strands. These cells move to lie between Reichert's membrane and the decidua and may help to establish and maintain a close relation between these two. A third group of less dense giant cells appear at about 7.0 day may be binucleate or multinucleate, and are found within the decidua. They are distinguished by their hyper chromatic nuclei, and are known collectively as the symplasia.

The decidua is entirely maternal, at least at the beginning. Dorsal to the embryo, toward the ectoplacental cone, is the decidua basalis which is not easily distinguishable from the uterine mucosa. Between the decidua basalis and the implanted embryo there will appear the richly vascular zone where the blood vessels appear actually to be ruptured. They appear more as sinuses than as formed blood vessels. Close around the embryo are the giant cells and ventral to it the decidua capsularis which is characterized by bi and possibly multinucleate giant cells in abundance. The decidua capsularis lies toward the newly establishing uterine-lumen, the side of the implantation which will come to lie freely within the newly established uterine cavity. It is directly opposite to the placental connections of the decidua basalis.

The nutrition must reach the embryo by way of the maternal blood. This vascular zone, between the embryo and the decidua basalis, consists of ruptured maternal vessels which loathe the entire embryo in blood, continuously changing. This may be seen first at about 5.5 days. The blood cells of the mother never actually penetrate the circulatory system of the embryo or fetus, being separated by Reichert's membrane, which is rather tough, and by the yolk cavity and the yolk-sac, thus, while Reichert's membrane is not living, it functions as semi-permeable membrane allowing soluble nutrients to pass through to the embryonic membranes and its blood vessels, particularly those of the so-called yolk sac. As the embryonic circulation develops, these nutrients are then circulated throughout to the developing tissues. The vascular zone expands through the decidual zone to the uterine lumen so that there may be some rupture of maternal blood into this lumen, beginning about 7.5 days and continuing for several days. The ultimate placenta is then made up of the decidua basalis, remains of the ectoplacental cone, the chorion and parts of the allantois forming a most efficient channel for nutrition from mother to embryo.

The development of the embryo from this stage onward is sufficiently like that of other vertebrates not to require detailed description. Its general shape at 8 days is not the typical figure "C" but rather a reversed capital "S" when observed from the left side or in sagittal section. The ectoderm is on the inside with neural groove open dorsally. Eventually the germ layers will be seen to be in their correct relationship to each other by a sort of reversed folding of the parts, accompanied by marked cephalization, the amnion, along with the chorion and yolk-sac, form a covering of the embryo marking its dorsal aspect. The "S" contour of the 8 day embryo gradually, changes and by 9.0 days the embryo has turned so as to present the capital "C" shape so characteristic of the higher vertebrates. The turning begins at both the cephalic and caudal ends and finally the mid-gut is involved at a time quite simultaneous' with the closing of the neural folds and the appearance of the notochord. Naturally the rotation of either end must be in the same direction, and is always toward the left side of the embryo. In sections at 8.0-8½ days this may be confusing because within the same field may be sections of the cephalic and of the caudal ends, each only apparently twisting in opposite directions. However, when the embryonic axis is considered, it is obvious that the entire embryo is turning in the same direction. The mid-gut of the Mouse is ill-defined, largely because in its early development the archenteron is an inverted groove. With the turning of the embryo' and the growth of the ends, the closure of the fore and hind gut, the lateral folds (splanchnopleure) connecting the two eventually come together to form a continuous tube by 9 days. The embryo comes to lie on its left side, close to the yolk sac and its right side is turned toward the opposite placenta. (Fig. 26.4)

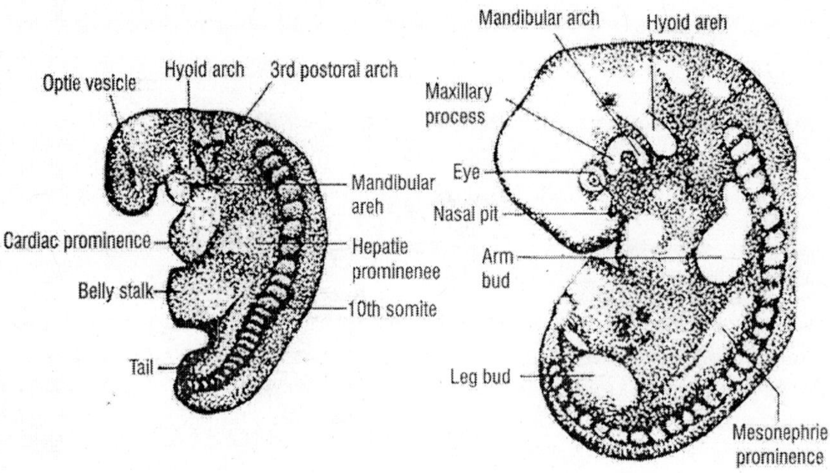

Fig. 26.4: 9 days old mouse embryo.

The mouse heart develops from the splanchnopleuric mesoderm, after the initial appearance of the cardiac coelom (pericardial cavity) as a split between the double layered splanchnopleure and somatopleure. The mesodermal portion of the splanchnotjture will give rise to both the covering and the muscles of the heart (epicardium and myoardium), originally known as the epi-myocardium. Other mesoderm, lying between the epi-myocardium and the endoderm will give rise by vacuolization to the pericardial cavity (originally part of the coelom). Unlike the development of the chick heart, where it is formed from two approaching masses of mesoderm from the sides, the mouse heart parts have a capital "U" shape with the base of the formation appearing first just anterior to the anterior intestinal portal and-ventral to the foregut and head fold. As the gut closes, the parts of-the "U" come together to form the single, tubular heart. Considerable cardiac differentiation may be seen by 9 days. Simultaneously with the development of the heart the major arteries appear, the dorsal aortae extending the length of the body from the paired aortic arches. Posteriorly they fuse to form the single omphalo mesenteric artery which passes out into the yolk sac where, at 8½ -9.0 days, it may disappeai as, a formed vessel and be seen as blood islands. But shortly the omphalo mesenteric veins, as in the chick, appear and return the yolk sac blood to the embryo by a more ventral route. While the yolk itself is so rudimentary it cannot supply nutrition for long, the yolk sac is a means of close association with the maternal tissues (vascular decidus) so that by this early circulatory relationship the embryo shortly begins to acquire nutriment from the mother.

Beginning at 9½ days one can see differentiated tissue areas and as development proceeds the mouse embryo becomes difficult to distinguish from the rat, pig, human or other mammalian embryos. Its size and availability make it of increasing value for the understanding of mammalian embryology. (Fig. 26.5)

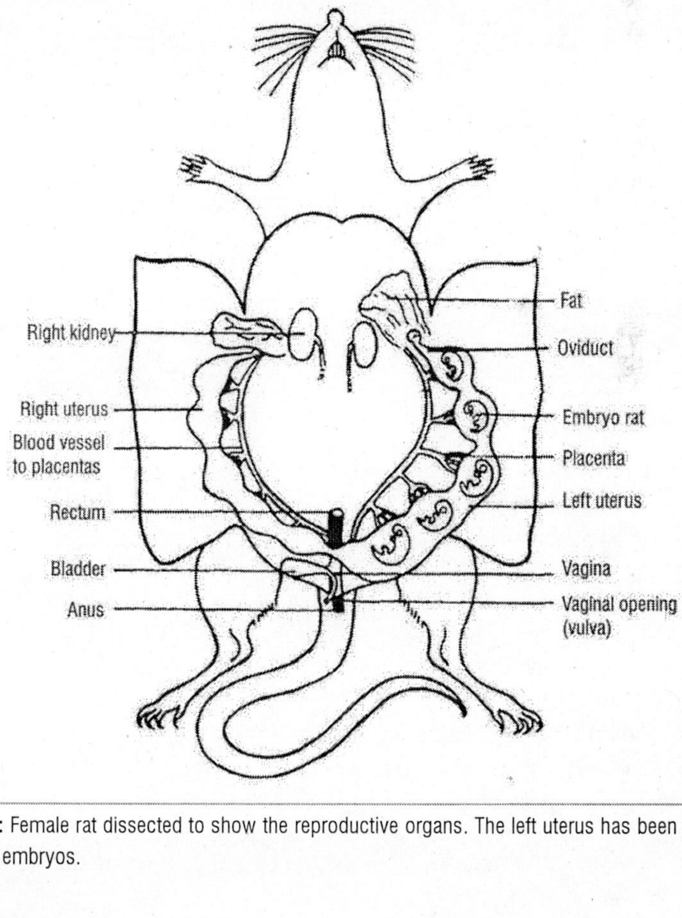

Fig. 26.5: Female rat dissected to show the reproductive organs. The left uterus has been opened to show the embryos.

CHAPTER 27

Chimaeras Development

Chimaeras are composite animals in which the different cell populations are derived from more than one zygote, this is in contrast to mosaics which are individuals derived from a single egg. Chimaeras can be produced experimentally by mixing two or more cell populations at a very early stage of development or by combining tissues from two or more individuals after the period of organogenesis.

The first type of chimaera is called a 'primary chimaera', the latter a secondary chimaera'. Primary chimaeras can be produced by two ways:

1. Cleaving embryos usually at about the 8-cell stage, can be aggregated in vitro so that they can meld into one double sized embryo. The resulting composite animals are called aggregation chimaeras (Fig. 27.1).
2. One or a few pluripotent cells can be injected into the cavity of the blastocyst. There they may intermingle with the cells of the inner cell mass and take part in the formation of the egg cylinder and the foetus. This can be produced by either using early embryonic cells, or by using teratocarcinoma cells. Resulting chimaeras are called injection chimaeras (Fig.27.2). Recently, chimaeras have been produced by aggregating pluripotent cells with cleaving embryos. Primary chimaeras have proven to be uniquely suitable experimental systems to trace the origin and fate of cell lineages in development. They also have offered rich material to investigate the interaction of two genetically different cell populations in the adult animal.

Procedure

Embryo Aggregation

1. Embryos at about the 8-cell stage, suitable for aggregation, are recovered from the uterotubal junction of pregnant females on day 3 of development. Some embryos may, however, already have entered the uterus. Therefore, the uteri are flushed in a separate dish too.
2. All embryos of one strain are collected, washed and transferred to a solution of pronase to remove the zona pellucida. The pronase solution is prepared by dissolving 10 mg of powder (Calbiochem) in 3 ml of Hanks' BSS in a small plastic Petri dish about 1 h before use.
3. The dish is pre-incubated at 37°C in an atmosphere of 5% CO_2 in air.
4. The embryos in the pronase solution are carefully observed under the dissecting microscope and occasionally aspirated into a fine capillary. As soon as the first few embryos lose their zone (usually after 3-4 min) all of them are quickly transferred into normal pre-equilibrated culture medium.

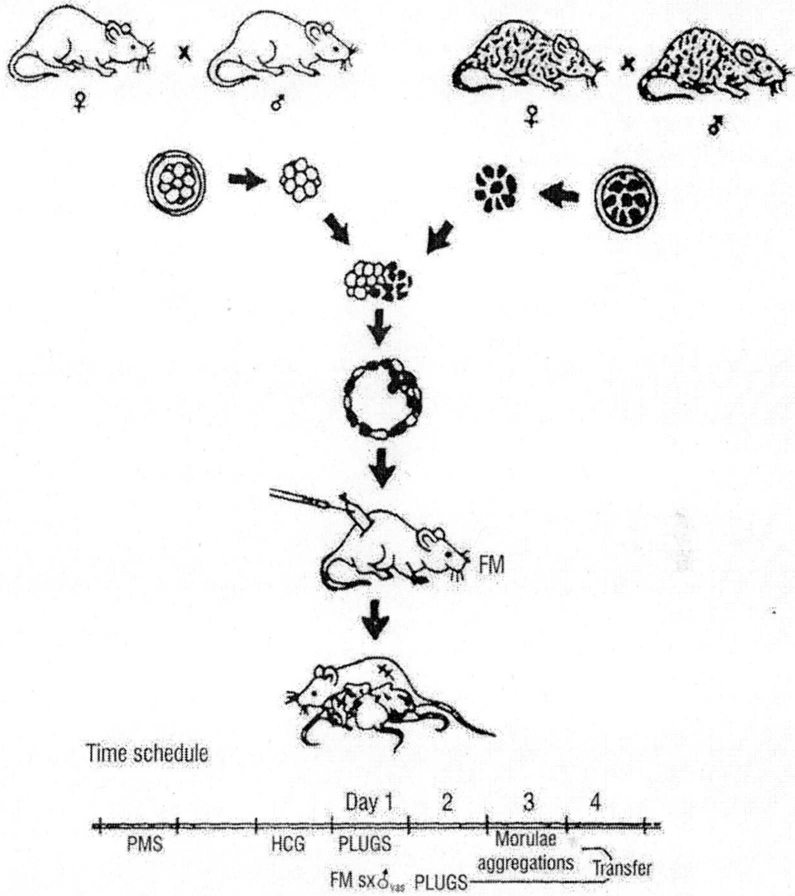

Fig. 27.1: Schematic outline for the production of aggregation chimaeras.

5. All the zonas are removed by repeatedly aspirating the embryos into a fine capillary.
6. The denuded embryos are washed twice in fresh medium and incubated until further use.
7. The partner embryos of the second strain used for chimaera production are prepared by the same way. Alternatively, the zona pellucidae may be removed by a quick exposure (a few seconds) of the embryos to culture medium acidified to pH 2.5. Embryos treated in this way readily aggregate.
8. For aggregation, embryos are transferred into small drops (3ml of medium covered with paraffin oil. The drops (up to 16 per small dish) are also prepared and pre-equilibrated 1-2 hr before use.
9. Two embryos, one of each strain, are placed side to side in each microdrop by

means of a mouth controlled capillary. When the medium is kept at 37°C and the zona removal had been mild, the two embryos spontaneously stick together without further addition of helper ingredients to the medium.

10. The loose aggregates are checked again one and two hours later and separated embryos brought into contact again. The dishes are incubated at 37°C in 5% CO_2 in air.

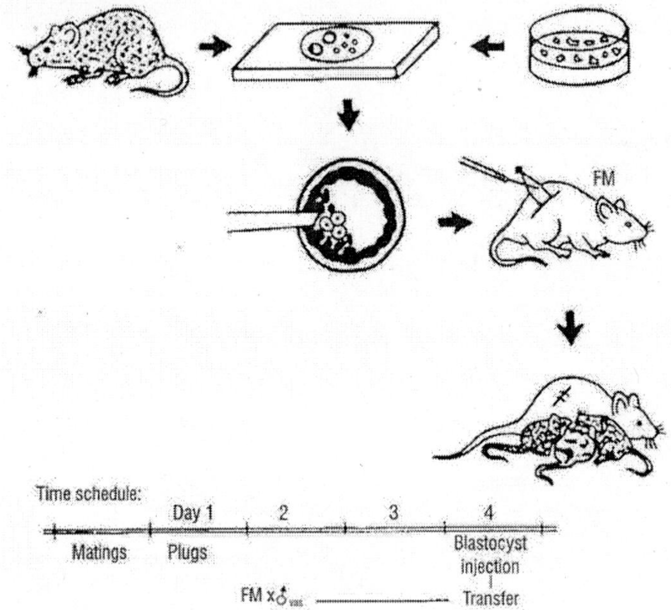

Fig. 27.2: Schematic outline for the production for the injection chimaeras.

11. The next day most aggregates (usually more than 90%) will have formed double-sized morulae or early blastocysts. Morphologically normal embryos are collected and transferred into the uterine horns of pseudo-pregnant foster mothers which are on day 3 of their pregnancy. About 10 aggregates are transferred into each uterine horn.

Cell Injection into Blastocysts

1. Blastocysts are obtained by flushing the uteri of previously mated females on day 4 of pregnancy. Not yet fully expanded, blastocysts are most suitable for injection, as fully expanded blastocysts tend to collapse immediately after puncture of the trophectoderm with the injection pipette.
2. Holding pipettes are prepared the same way as the ones used for eggs. Injection pipettes are pulled from borosilicate tubing (1.2 mm) to a capillary diameter of about 20mm (depending on the size of the cells to be injected). The tip of the capillary is first broken and subsequently polished on a grinding

wheel.

3. Ideally, the opening should form an angle of 30-40°. The injection pipette is finished by drawing a fine, sharply pointed tip and by slightly closing the opening. This is achieved by melting the tip of the pipette to a hot filament. As soon as the borders of the opening start to melt and slightly collapse the pipette is quickly withdrawn. Before use the injection pipettes are siliconized.
4. For injection the blastocysts together with the cells to be injected into the blastocoel are placed into a drop of medium covered with paraffin, oil. Preferentially, the cell culture medium is used. Injections are executed at low magnification using the inverted microscope equipped with two manipulators.

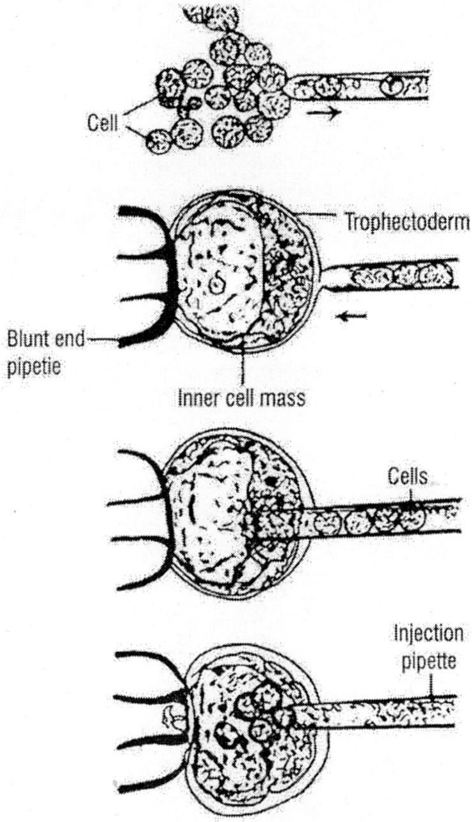

Fig. 27.3: Cell injection into blastocyst. Cells are sucked into an injection pipette (a). A blastocyst is immobilized at the tip of a blunt holding pipette and an injection pipette containing cells is set to puncture the trophectoderm and the inner cell mass (b). After penetration of the injection pipette into the blastocoel (c), the cells are deposited onto the inner cell mass (d) and the injection pipette is withdrawn.

5. For cell injection; both pipettes, the holding pipette and the injection pipette,

are partially prefilled with medium. A cell injection is illustrated. (Fig. 27.3.)

6. One blastocyst after the other is immobilized by suction to the holding pipette. Usually, the inner cell mass is directed towards the opening of the injection pipette. The cell(s) to be injected are aspirated into the tip of the injection pipette. The tip of the injection pipette is directed to an intercellular junction of the trophectoderm and the cell layer is penetrated with a rather slow move.
7. The cells are deposited onto the inner cell mass and the pipette withdrawn from the blastocoelic cavity. The injected blastocysts usually collapse; however, they reform a blastocoelic cavity when incubated for a few hours. After recovery they are transferred into the uterine horns of day 3 pseudo-pregnant foster mothers. About 10 injected blastocysts are transferred into each uterine horn.

Cell Injection into Post-Implantation Stages

1. Post-implantation stages at around day 10 of development may be injected with foreign stem cells across the uterine wall. At this stage especially the placenta can be relatively easily be located by superficial examination. However, to become acquainted with the localization of placenta and fetus inside the uterine swellings a few implants of the corresponding stage should be dissected before injecting cells into the living organism. The uterine horns of pregnant females are exposed across dorsolateral incisions.
2. Cells are injected by means of finely drawn capillaries of about 20-30mm in diameter
3. Connected to a micrometer syringe.
4. The tip of the capillaries is broken with watchmaker's forceps to produce a sharp-pointed bevel and siliconized before use.
5. About 1-2 x 105 viable fetal liver cells may be injected into the placenta to prevent genetic anemias (Fleischman and Mintz, 1979). Injection of about 50-300 neural crest cells are sufficient to produce chimaerism of the melanocyte population.

Establishment of Embryonic Stem Cell Lines

1. Embryonic stem cell lines are derived from the proliferating inner cell masses of in vitro outgrowing blastocysts. For outgrowth unhatched blastocysts on day 4 or hatched blastocysts early on day 5 of development are flushed from the uteri of spontaneously mated females.
2. Groups of 5 blastocysts are placed into paraffin oil covered microdrops (20mm) of DMEM medium supplemented with 20% fetal calf serum in plastic tissue culture dishes.
3. Several batches of serum should be compared with regard to the support of optimal outgrowth.

4. After 5-6 days in culture (37°C, 5% .CO_2 in air) large inner cell masses are picked off the outgrown trophectoderm by means of a finely drawn, mouth-controlled capillary and partially dissociated using trypsin/EDTA.
5. The resulting cell clumps are carefully washed and seeded at high density onto a feeder layer of mitomycin C-inactivated STP fibroblasts cultured in 6-nuii wells of 96-well plates (medium: DMEM + 10% FCS).
6. Growing colonies of embryonic stem cells are picked off individually, partly dissociated and subcultured at approximately 7-day intervals.
7. For injection, colonies of embryonic stem cells are picked off the culture dishes and trypsinized. Single cell suspensions are incubated in small tissue culture dishes for 1-2h.
8. After this incubation period feeder fibroblasts have reattached to the bottom of the dish.
9. Nonattached cells are collected, concentrated by centrifugation and added to microdrops containing blastocysts prepared for injection. Up to 8-12 cells may be injected into the blastocoel of each blastocyst.

CHAPTER 28

Study of Embryological Slides

Development of Frog

Fig. 28.1a Left: A section through a frog blastula. Note the small cells in the animal half the micromeres; the large cells containing yolk in the vegetal half the macromeres; the cavity blastocoel-normally filled with a fluid. A dark brown pigment is present in the periphery of the egg at the animal pole and extends over the surface of the egg to a parallel just below the equator. The vegetal pole and adjacent regions are white. The blastula lies within a thin vitelline membrane. **Right:** A section through the very early gastrula. Note the increase in the number of cells above the blastula and the appearance of a heavily pigmented notch, the blastopore. The region above the blastopore invaginates and at the same time grows over the yolk.

Fig. 28.2a Left: A section through gastrula of the frog. Note the invagination of the dorsal lip of the blastopore marked by the pigment in the surface of the dorsal lip; the narrow canal is the archenteron; the yolk is beginning to fold over to enclose the blastocoel. The dorsal lip is moving over the surface of the yolk. **Right:** A dissection of a gastrula showing the blastocoel, the heavily pigmented animal hemisphere. The sharp, crescent shaped line between the black and white hemisphere is the dorsal lip of the blastopore. This lip is moving over the surface of the white hemisphere.

Fig., 28.3a. Left: A frog gastrula with a dorsal and ventral lip, indicating that the blastopore is now circular. A plug/ of yolk stoppers the blastopore; the archenteron is increasing in size at this time, but the blastocoel is decreasing. By this time the region which was originally above the dorsal lip at the beginning of gastrulation has completely invaginated to form the roof of the archenteron. Above the roof of the archenteron is the presumptive neural plate. **Right:** A dissection of a late gastrula, showing the large archenteron. The small blastocoel is embedded in the yolk mass and does not show. Note the turning in of cells at the ventral lip to form some of the mesoderm.

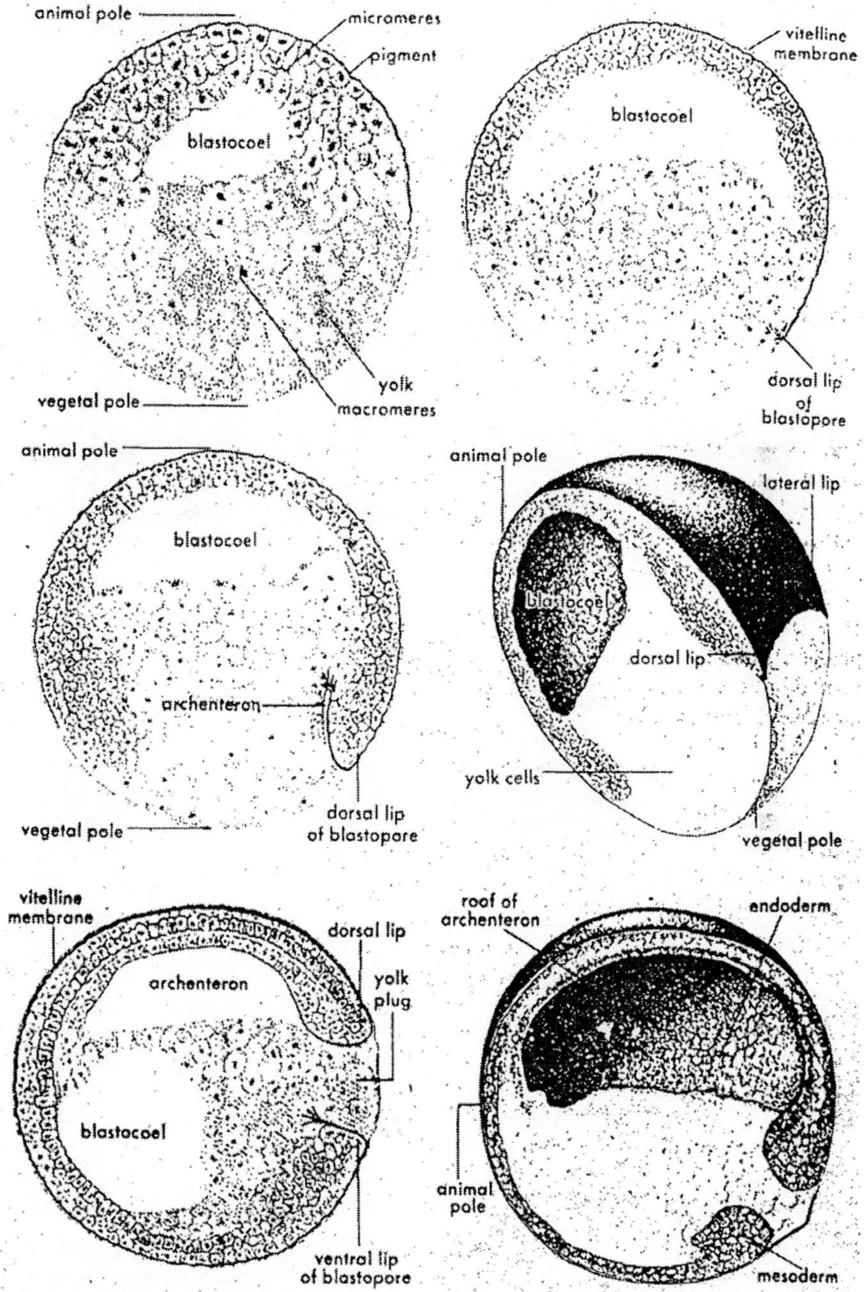
animal pole
micromeres
pigment
blastocoel
yolk
vegetal pole
macromeres
vitelline membrane
blastocoel
dorsal lip of blastopore
animal pole
blastocoel
archenteron
vegetal pole
dorsal lip of blastopore
animal pole
lateral lip
blastocoel
dorsal lip
yolk cells
vegetal pole
vitelline membrane
dorsal lip
archenteron
yolk plug
blastocoel
ventral lip of blastopore
roof of archenteron
endoderm
animal pole
mesoderm

Fig. 28.1b: The end of gastrulation in the frog, showing a small blastopore and the yolk plug in the process of being withdrawn. The ectoderm is two cells in thickness. The endoderm is still in the form of large, yolk laden cells. The notochord is part of the roof of the archenteron at this time, but it will soon split free and the endoderm will then migrate to form the archenteric roof, resulting in a true gut. The mesoderm extends around the lateral sides of the gut and can be seen near the ventral lip of the blastopore. No blastocoel is seen at this time. The blastopore marks the posterior end of the future embryo. About 50,000 cells compose the gastrula at this time. The relation of the gastrula to the future embryo may be seen by comparing this figure with figures 28.2c.

Fig. 28.2b: The neural plate stage of the frog. Now the ectoderm cells above the notochord appear elongate and form a neural plate, which later folds to form the neural tube. The anterior end of the plate, which is thick, is the brain region. Underneath the neural plate lies the notochord, which is now distinct and separate from the endoderm of the roof of the gut. The yolk plug has withdrawn and the blastopore lips met. A comparison with figures 28.2c and 28.2e illustrates the subsequent changes in the neural plate and in the anterior end of the gut. Note the increasing complexity developing in the walls of the brain; the evagination of the pharynx to fuse with the ectoderm, forming the oral plate; the outgrowth of the tail bud to form the tail; the differentiation of the solid notochord cell into a vesicular type of cell, and the appearance of mesodermal somites.

Fig. 28.3b. This figures of the frog illustrate the formation of the neural tube from the neural plate and the separation of the lateral mesoderm from the endoderm. **Left:** The notochord and lateral mesoderm are distinct from the endoderm and in close contact with the neural plate. Transplantation experiments show that the notochord and lateral mesoderm induce the neural plate to form from the ectoderm. Note that the ectoderm is still two layers thick an outer epidermal layer and an inner sensory layer, which is continuous with the thick neural plate. **Right:** The neural plate has folded into a plough which will fuse dorsally to form a tube. This action takes place inside the epidermal layer of ectoderm. Further development of the neural tube may be seen in figure 28.4d, where the tube is completely closed, and in figures 28.3f where internal differentiation of the tube has taken place.

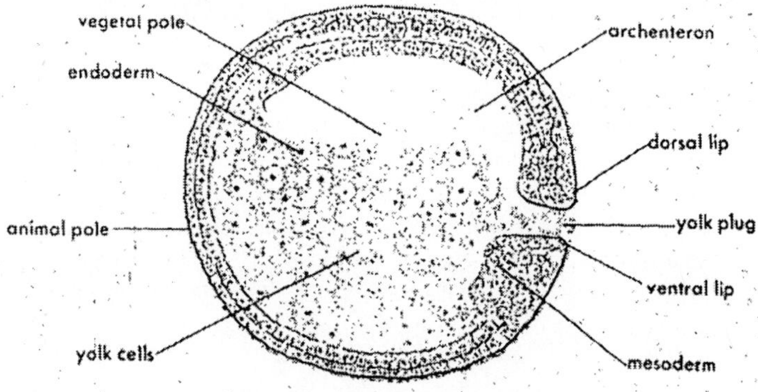

28.1b

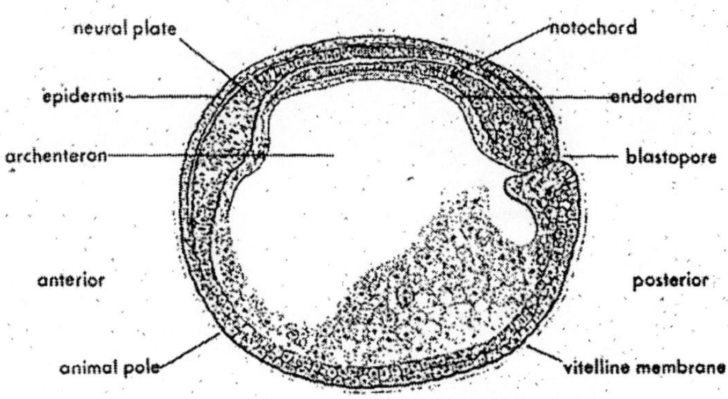

28.2b

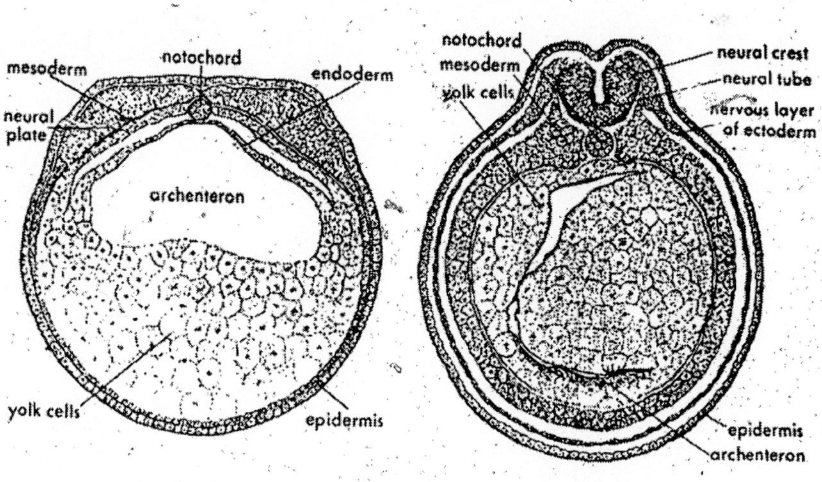

28.3b

Fig. 28.4b: A longitudinal section through the late frog neurula. A comparison with figures 28.2c and 28.2e will furnish the intermediate stages between the neurula and the swimming tadpole. Note a group of cells from the nervous layer of the ectoderm just underneath the brain, the primordium of the anterior part of the pituitary gland; a slight depression of the epidermal layer of the ectoderm, the stomoderm; a short diverticulum of the pharynx, the liver evagination. Posteriorly the neural canal is continued into the gut as a neurenteric canal, and a proctodeum is present, but these structures are not in the section illustrated here.

Fig. 28.1c: Sections showing the closure of the neural folds in the frog to form a neural tube and the fusion of the endoderm beneath the notochord. Note also the complete band of mesoderm between, the ectoderm and the endoderm. This band is composed of two layers of cells which, when separated, enclose a cavity, the coelom. The mesoderm extends dorsally on either side of the neural tube. The thickened portion of the mesoderm adjacent to the notochord is nephrotome and, as seen in figures 28.4d and 28.3f, becomes organized into the pronephros. It reveals the changes in the neural tube, the notochord, the gut, and the liver.

Fig. 28.2c: A dissection of a tail bud embryo. The neural tube is beginning to show the three primary divisions of the brain, with the optic vesicles developing from the prosecephalon. The dorsal lateral mesoderm has formed into somites. In the endoderm we may distinguish the pharynx, liver diverticulum, the hindgut, and the anus. The heart is still in the process of organization from the ventral mesoderm.

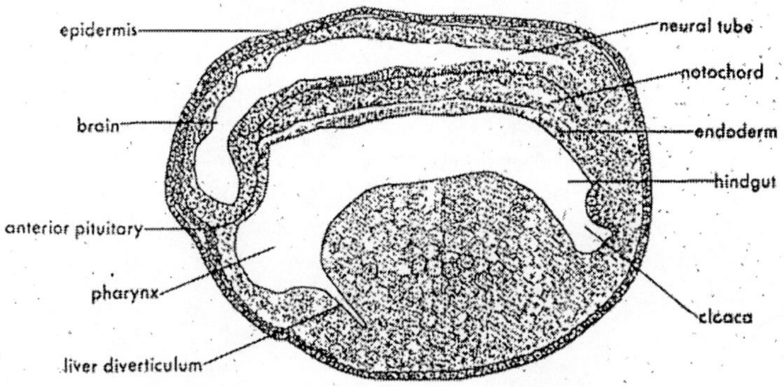

28.4b

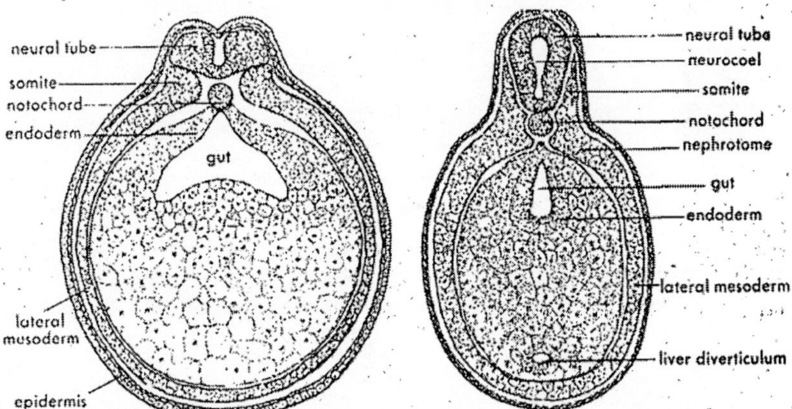

28.1c

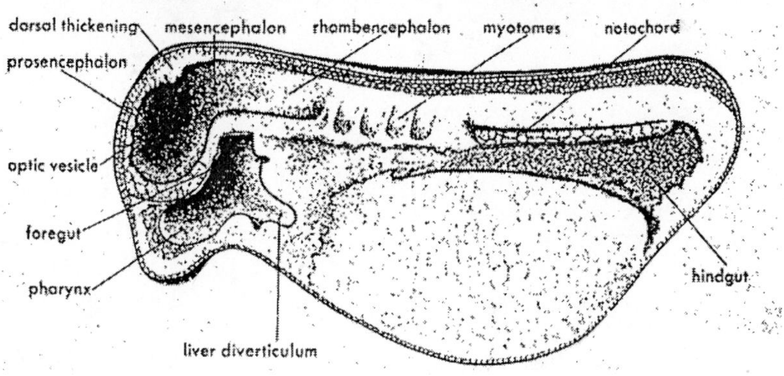

28.2c

Fig. 28.3c: A section through the very early optic vesicle of the frog (left) and one through a later stage (right), when the lens is beginning to form within the optic cup. The eye Begins as a simple outgrowth from the forebrain. When the optic vesicle touches the inner layer of ectoderm, a lens is induced. The vesicle then becomes cup-shaped, and its connection with the brain narrows to a thin stalk, the optic stalk. Next, the thick layer of the cu adjacent to the lens differentiates on a fibrous structure and finally becomes transparent. The ectoderm and mesoderm near the lens also become transpazent, forming a cornea. The optic nerve is made up of nerve fibers, which originate in the retina and migrate by means of the optic stalk' to the brain.

Fig. 28Ac: A section through the ear region of a 4 mm, frog embryo. The otic vesicles arise from otic placodes in the ectoderm opposite the hind brain. In addition, figure. 28.2f shows the further development of the internal ear in the frog. After separating from the ectoderm, the otic vesicle transforms, fonning the endolymphatic duct, the semicircular canals, and the sensory organ for sound. Nerves develop adjacent ganglia and connect the internal ear with the brain.

Fig. 28.1d: A longitudinal section of a 6 mm. frog embryo. This is shown to present an over all picture of this stage of development in preparation for the following series of cross sections, which show details of structure at higher magnification. By this time the brain has undergone considerable differentiation, and the epiphysis, the infundibulum, and the optic chiasma an distinct. The endoderm continues to develop into liver diverticulurn, and oral evagination an indications of from thyroid are apparent. A well-developed tail with mesodermal somites has differentiated from the tail bud. A large mass of yolk still remains, serving as a food supply.

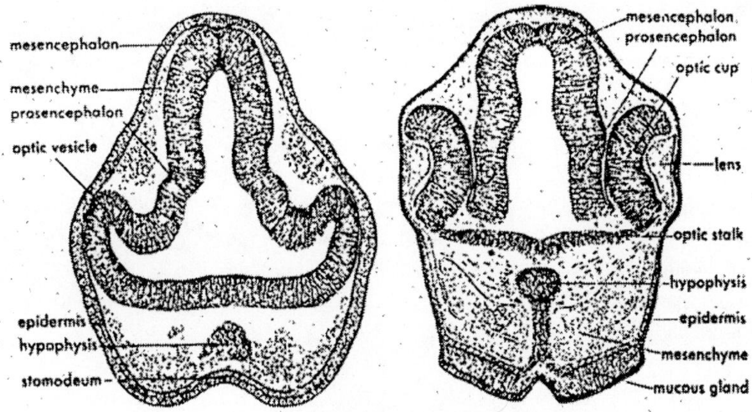

28.3c

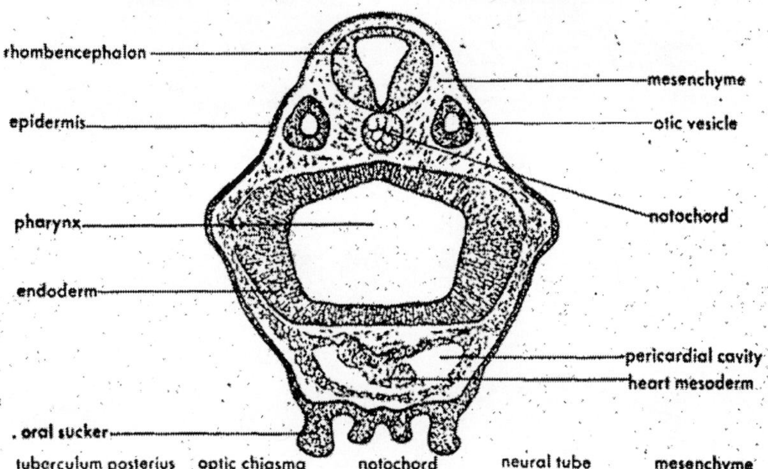

28.4c

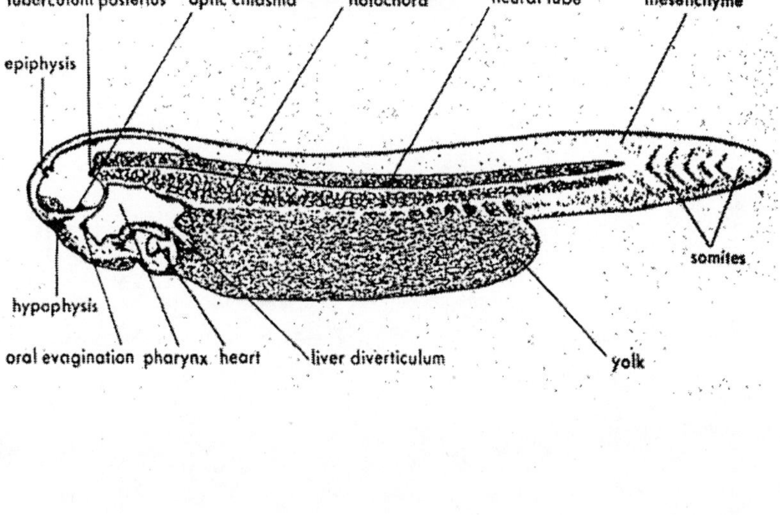

281d

Fig. 28.2d Left: A section through the nasal pits and epiphysis of a 6 mm. frog embryo. Thick nasal placodes are invaginating to form the pits. The invaginations become deeper as time goes on, and in figure 28.2e connect with the pharynx. The thicker part of the invagination forms the olfactory epithelium and sends sensory nerve fibers to the brain. The connection with the exterior remains open, forming the external pares. **Right:** A section through the eye. A well formed lens is present in the eye cup, and the eye stalk has narrowed to a thin strand of tissue. (compare with fig. 28.10. In figure 28.2d the hypophysis is now between the brain and pharynx. The thyroid develops from the floor of the pharynx and divides into two parts, as in figure 28.1f. The mucous glands are clearly differentiated and secrete mucus at this time.

Fig. 28.3d: A section through the otocyst, auditory vesicle, and otic vesicle of the frog. Dorsally the vesicle now shows a small bud, the endo lymphatic duct. The auditory nerve 8 develops from part of the early auditory placode, which also gives rise to the seventh facial nerve. The main body of the auditory vesicle becomes divided into saccule and utricle. This figure also shows the conus arteriosus leading into the afferent arteries, which join with efficient arteries leading to the dorsal aortae. Gill filaments have developed, and blood will soon circulate through the gills.

Fig. 28.4d: Sections through the early pronephros of the frog. The nephrotome forms tubules which connect with the coelom by means of three nephrostomes. A glomus develops in relation to the dorsal aorta, and a pronephric duct conducts excretions to the cloaca. The tubules become twisted, as in figure 28.3f, and the nephrostomes become reduced from three to two. Note also the liver diverticulum, the vitelline veins, and the mesodermal somites on either side of the notochord. The mesoderm has split into a somatic and splanchnic layer enclosing a coelom, and the somites have separated from the nephrotome. The gut, which is still composed of large, yolk-laden cells; is well organized into esophagus, lung buds, and stomach.

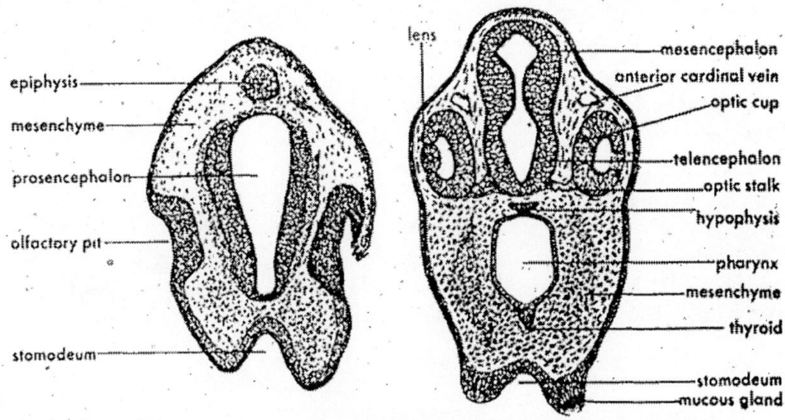

28.2d

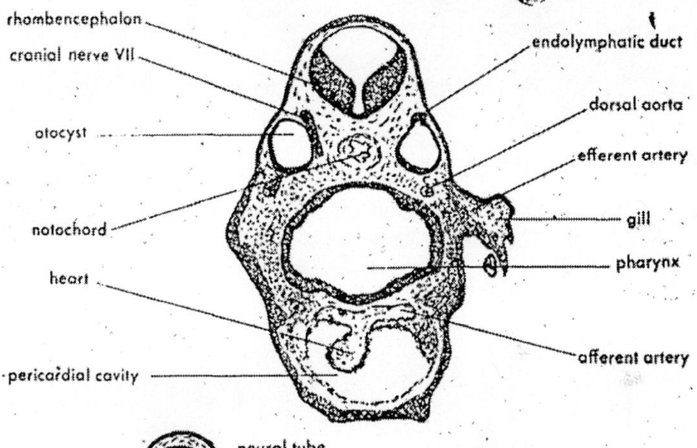

28.3d

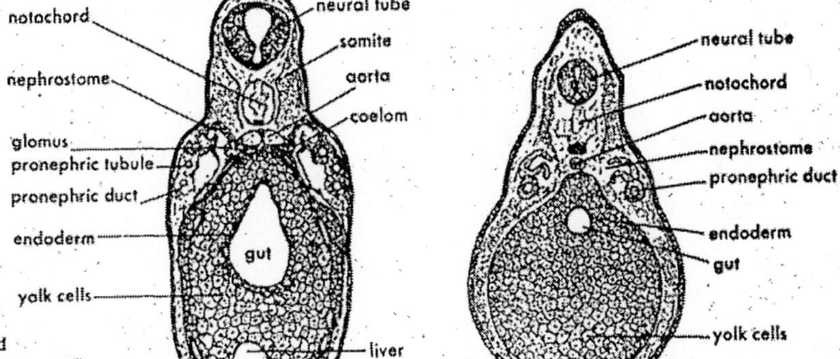

28.4d

Fig. 28.1e: Sections through the cloaca and anus of the 6 mm. frog embryo. The pronephric duct may be seen entering the cloaca on either side of the midline. In this figure the somites are differentiating into muscle fibers, and in figure 28.4f the muscles are flinctional. Note the disappearance of the yolk between the 6 mm. embryo and 10 mm. tadpole. In figure 28.1c note also the changes in the notochord from a rod of ordinary appearing cells to a relatively swollen tube of vesicular cells. The notochord serves as a foundation for the formation of the vertebrae in later development.

Fig. 28.2e: A longitudinal section of an 8 mm. frog embryo. This bridges the gap between the 6 mm. embryo and the 10 mm. embryo. The 8 mm. larva has hatched and exhibits active swimming movements. The heart beats, circulation is well established through the gills, and the mouth is about to open by the breaking of the oral plate. Food is as yet unnecessary, as evidenced by the large yolk mass. The brain walls continue to differentiate, and five primary divisions are now discernible. Certain land marks-the optic recess, infundibulum, and tuberculum posterius are present.

Fig. 28.3e: The 10 mm. frog larva in longitudinal section. This figure is introduced to show an overall picture of this stage; the following sections will show details of structure. Generally speaking, this is a complete functional larva. The mouth is open and the larva feeds. Note that the mass of yolk has been used up and that the coils of the intestine occupy the region of yolk. The notochord has become highly vacuolated, while the somites are now definitely muscular, forming parallel bands running out into the tail. The brain is too complicated for a study at this low magnification and will be shown in the following sections.

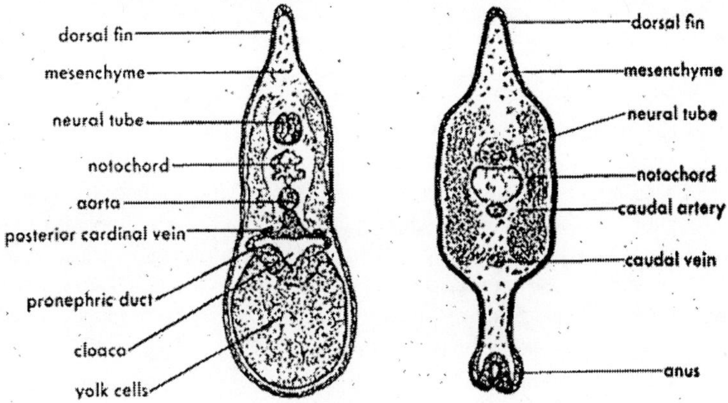

28.1e

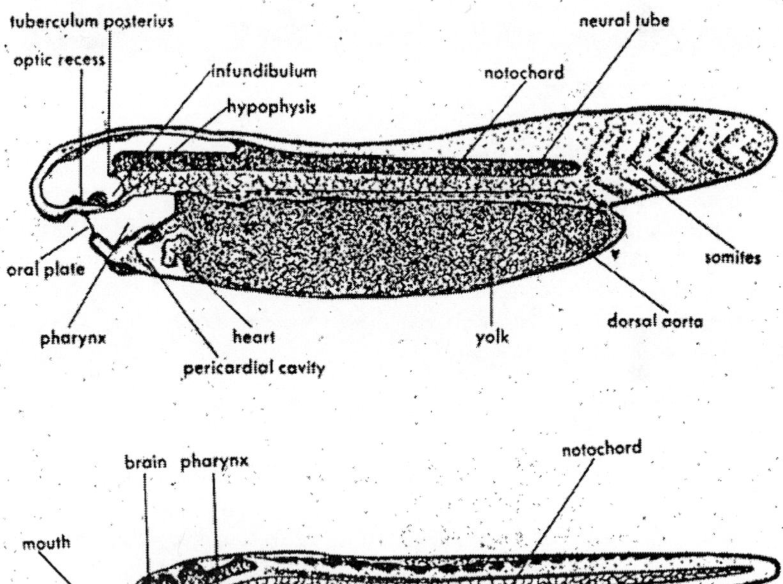

28.2e

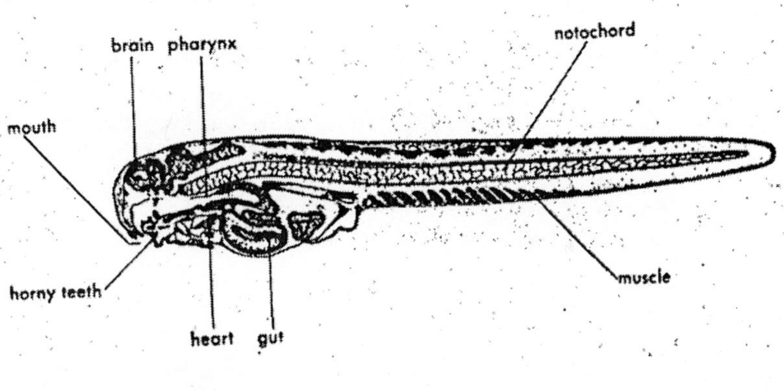

283e

Fig. 28.4e: A section through the olfactory organ of a 10 mm. tadpole. The olfactory tube opens to the exterior through the external nares just a fraction of a millimetre anterior to the section shown. The olfactory epithelium is the portion which becomes sensory and connects with the brain by means of the first cranial nerve, the olfactory nerve. Since the larva is free swimming at this time, the mucous glands no longer serve any function; they will soon disappear completely. Notice the change in the head mesenchyme in this section, as compared" with figure 28.-2d. Mesenchyme cells have aggregated in clumps to form centers about which cartilage develops. The relatively large cartilage cells embedded in the cartilaginous matrix may be seen below the oral cavity and adjacent to the olfactory organ. The remainder of the mesenchyme forms loose connective tissue which secretes a jellylike matrix.

Fig. 28.5e: A section through the olfactory organ of a 10 mm. tadpole (compare with fig. 28.2d). The olfactory tube opens into the pharynx through the internal flares. Dorsally and laterally, the olfactory epithelium is known as the lateral appendix, and adjacent to it is the lateral groove. The olfactory nerve has differentiated from the neuroblasts in the olfactory epithelium and enters the right side of the brain. The brain shows a differentiation into deep staining cell bodies and light-staining nerve fibers. Although there is a very close relationship between the olfactory organ and the telencephalon, transplantation experiments show clearly that the early nasal placode will differentiate into a normal olfactory organ in the absence of the telencephalon. Large masses of cartilage are located below the brain and to either side of the pharynx.

Fig. 28.1f: A section through the eye of a 10 mm. tadpole. The inner layer of the optic cup has differentiated into the sensory layer of the retina, while the outer layer has deposited pigment in order to exclude light. The sensory layer shows further division into a ganglionic layer of nerve cells and nerve fibers and into a layer of rods and cones, which are the actual transformers of light energy into nerve impulses. The lens has lost its cavity and is now composed of two main groups of cells, a thin epithelial layer and an internal mass of lens fibers. The latter become transparent by the formation of a hyaline substance. Next to the lens the cornea develops from the ectoderm and some mesenchyme. It, too is composed of an outer epithelial layer and an inner fibrous, but transparent layer. A fragment of the optic nerve may be seen between the left eye and the brain.

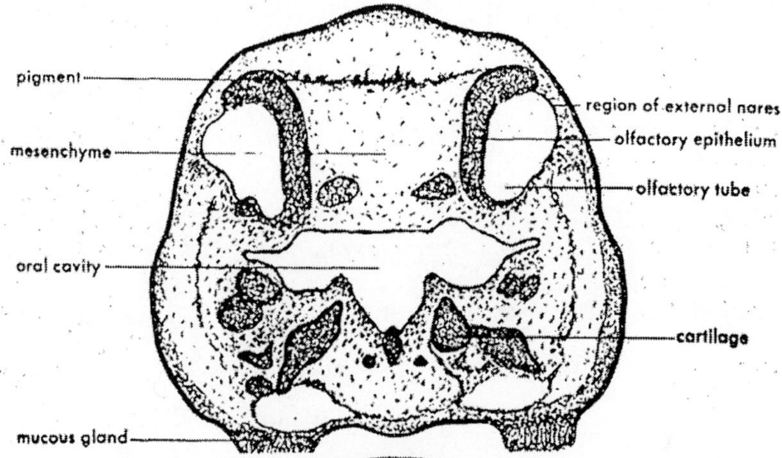

28.4c

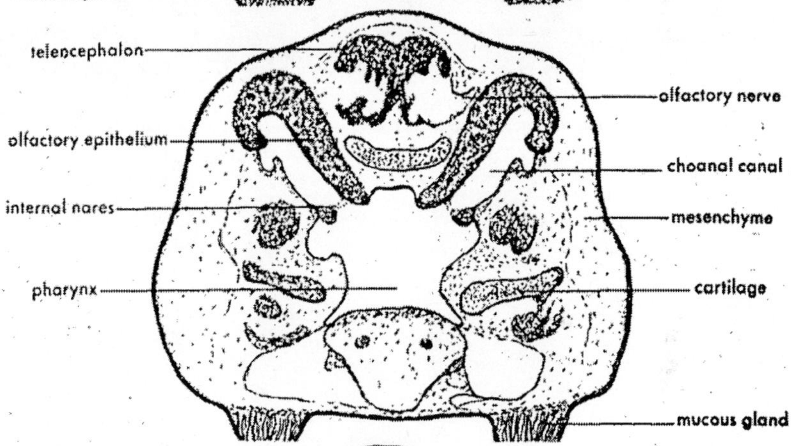

28.5e

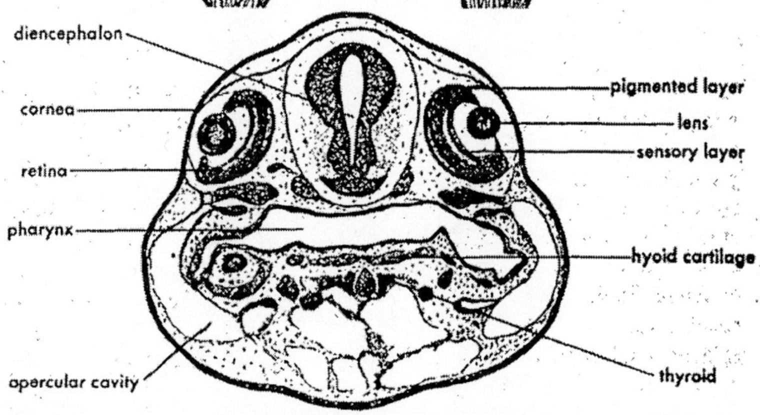

28.1f

Fig. 28.2f: A section through the internal ear of a 10 mm. tadpole. The early pear-shaped otic vesicle has differentiated into saccule, utricle, and semicircular canals. This section passes through the saccule and one of the canals. Between the ear and the hind-brain lie fragments of the eighth cranial ganglion. This ganglion develops from a part of the early auditory vesicle. Just ventral to the internal ear are the paired cardinal veins and paired aortae. The pharynx opens laterally into the opercular cavity, and water taken into the mouth is forced through the openings over the gills and finally out through a spiracle. The floor of the pharynx opens through a slit-the glottis into a tube which branches and leads back to the lungs. These branches are the trachea and bronchi. The drawing also shows the liver and two sections of the intestine.

Fig. 28.3f: A section through the pronephros of a 10 mm. tadpole. The first beginning of the pronephros is simply as a thickened layer of mesoderm between the somites and the lateral mesoderm. These cells rearrange themselves to form three pronephric tubules, which communicate with the coelom through nephrostomes. The three tubules fuse posteriorly, and a pronephric duct forms which leads back to the cloga. A vascular body the glomus develops, with connections to the dorsal aortae. The fully developed pronephros of figure 28.3f exhibits a coiling of the pronephric tubule and a consolidated glomus. The large irregularly shaped esophagus is flanked by the paired lungs. Well differentiated spinal cord, with its white fibers and gray nerve cell bodies.

Fig. 28Af: A section through the trunk of a 10 mm. tadpole. The drawing shows the cloaca and entrance of the pronephric duct. The latter will become the Wolffian duct of the mesonephres. Note that all of the visible yolk of figures 28.1c and 28.1e has now disappeared. The yolk granules have been broken down by enzyme action, and the digested products have entered the blood stream. Of particular interest is the differentiation of the muscle of the trunk. The cells of the somites of figure 28.1c have developed into myoblasts, which secrete inside themselves a contractile protoplasm in the form of fibers. The section cuts across the striated muscle cells of figure 28.4f, which are arranged in groups. The notochord is large and its cells have become vacuolated. The notochord forms a supporting rod in the absence of vertebrae. The epidermis and some supporting mesenchyme form a thin dorsal fin.

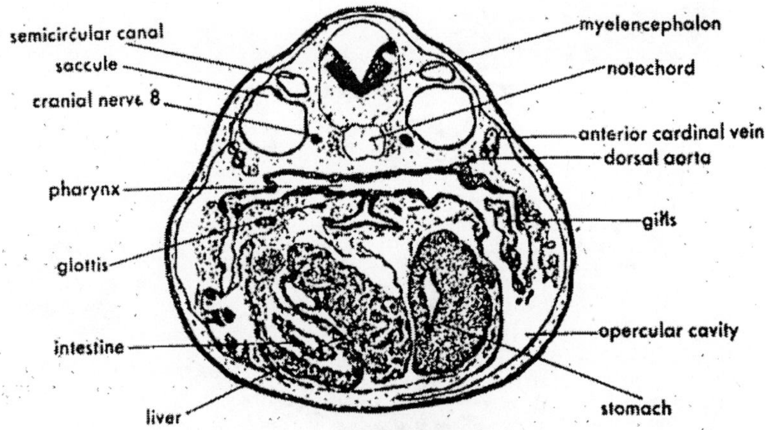

28.2f

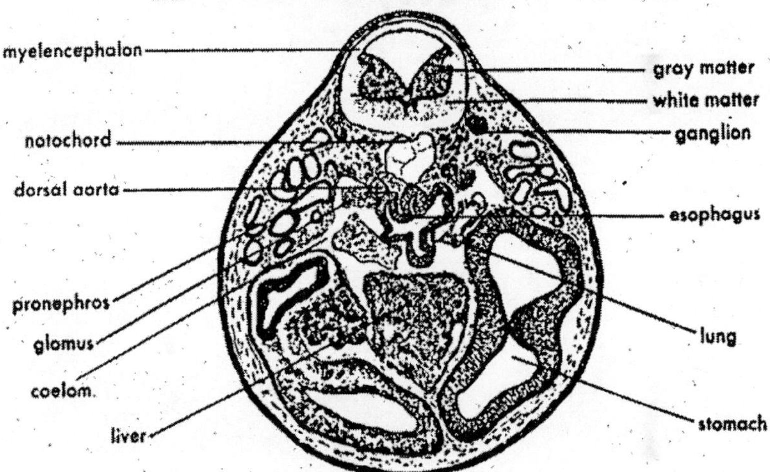

28.3f

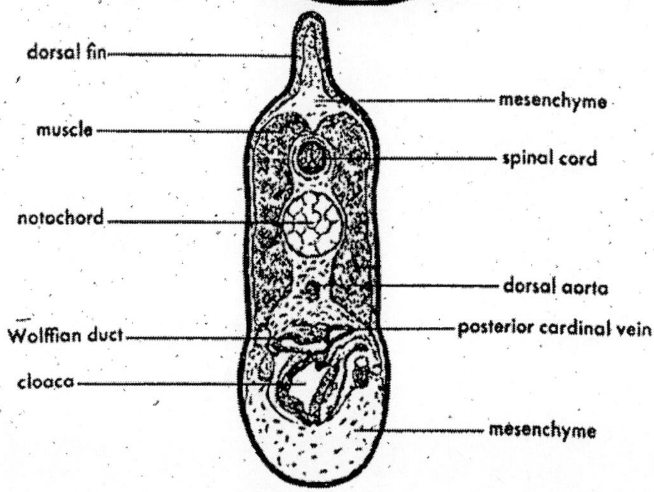

28.4f

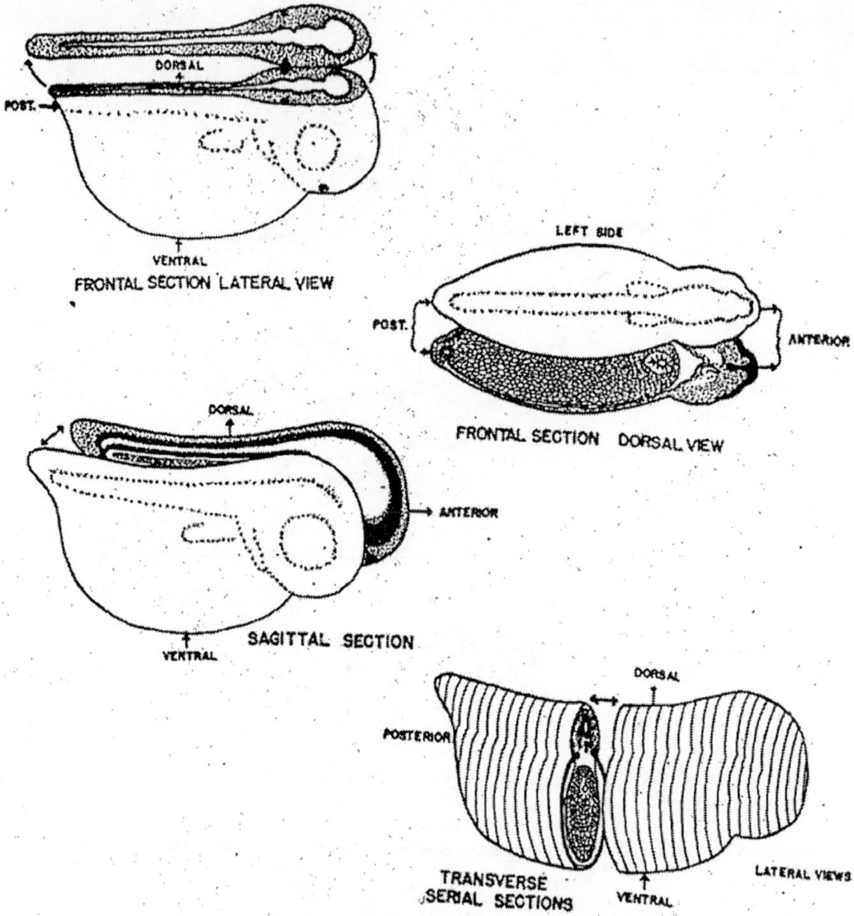

Diagram presenting the planes in which the embryo may be cut or sectioned.

The search for truth is in one way hard and in another way easy. For it is evident that no one can master fully nor miss it wholly. But each adds a little to our knowledge of Nature, and from all the facts assembled there arises a certain grandeur.

Aristotle

CHAPTER 29

Study of Embryological Slides

Development of Chick

Transverse Sections of 20 Hours Chick Embryo (Fig. 29.1. a, b, c and d)

With the structures of the primitive streak whole mount clearly in mind, study the serial transverse sections of an embryo of 18 to 20 hours of incubation with the object of being able to distinguish at all times between sections taken through the streak and through the embryo proper.

First identify the marginal regions where the blastoderm is in direct contact with the large yolk granules beneath. This is the region of the area opaca. Midway between two such areas locate the primitive folds within the area pellucida. The sections should be studied with the yolk ventrally placed, i.e., toward the observer as seen through the microscope. The most dorsal germ layer is ectoderm which shows considerable thickening toward the center of these transverse sections. The thickening may be solid (Hensen's node) or may be depressed into a groove (primitive groove) on either side of which are folds of ectoderm (primitive folds). It is of utmost importance that the student learn to distinguish between the primitive and the embryonic structures. There are, anterior to the streak, neural folds and a neural groove. There are several ways to identify the primitive streak in transverse section and to avoid confusion with the neural structures. In the first place there is no notochordal tissue beneath the primitive streak. In the second place the mesoderm of the primitive streak region is never organized into somites, is more concentrated at the streak, and the mesodermal cells seem to be coming directly from the streak (ectoderm) cells. In the third place the ectoderm, mesoderm, and ventral endoderm all merge at the streak while in sections through the embryo proper the germ layers are separate.

Lateral to the primitive folds locate the rather loosely scattered cells which seem to be migrating outward between the upper layers of ectoderm and the lower layer of very thin endoderm. These cells represent the mesoderm which originates from the primitive streak and it migrates anterior to take up positions related to the embryo itself. Both in the area opaca, where blood islands are to appear, and close to the neural folds where the somites will appear.

The space beneath the embryonic structures, cleared of yolk and lying below the thin endodermal layer, is known as the primitive gut. This cavity is continuous beneath both neural and primitive streak regions. The floor of this primitive gut consists of yolk which was left behind when the blastoderm was removed for mounting.

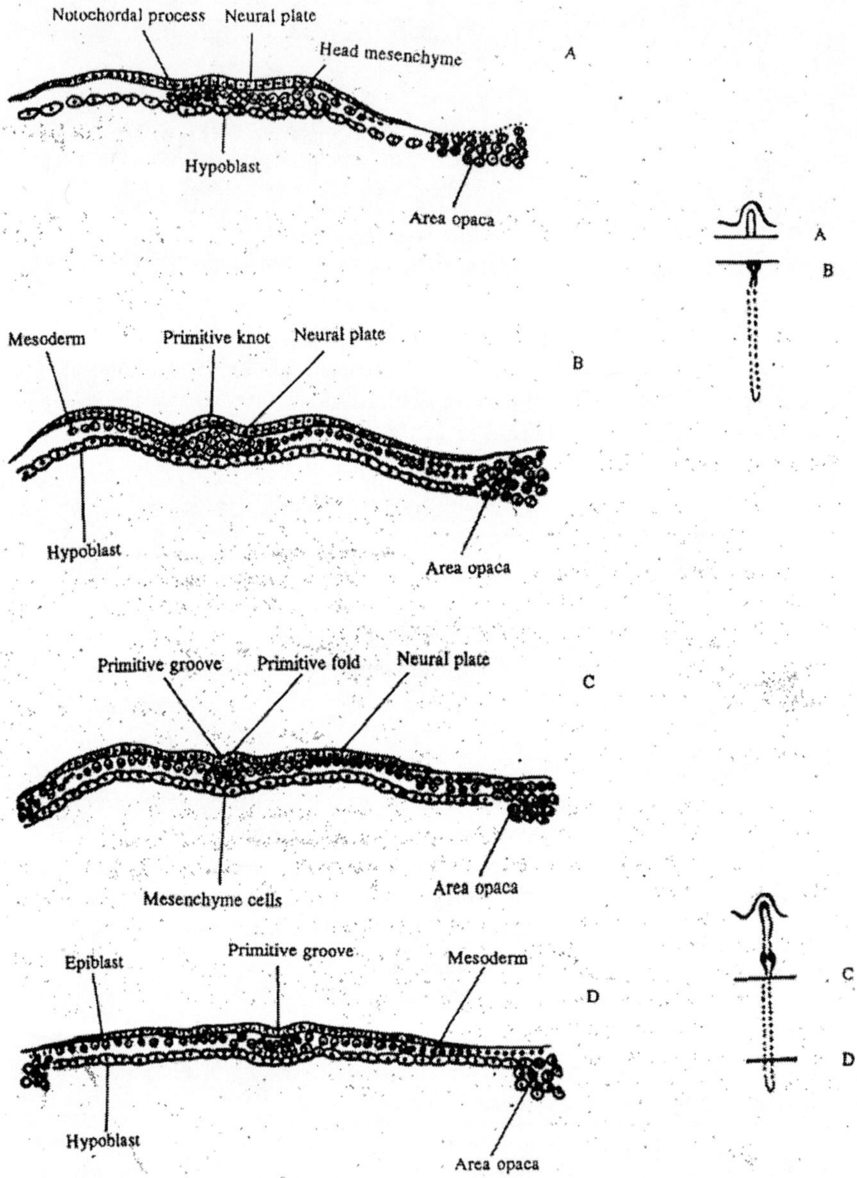

Transverse Sections of 24 Hours Chick Embryo (Fig. 29.2 a, b, c, d)

In studying the serial transverse sections of the 24 hour chick embryo the student should locate the most anterior sections first and then proceed to examine the series posteriorly, identifying each structure as it appears and disappears, making frequent reference to the whole mount photograph and indicating in this picture exactly where each detailed transverse section was taken. The structures seen in a typical series of sections through a 24 hour chick embryo will be described briefly in the order in which they should appear, from anterior to posterior.

The most anterior sections will show the anterior tips of the lateral neural folds which are not yet fused around the anterior neuropore. Directly below is the double-layered pro-amnion, made up of ectoderm and endoderm, and limited laterally by the adherent yolk cells of the marginal area opaca. There are cellular differences in the ectoderm and endoderm which make them always distinguishable. Locate the germ wall and more laterally, the earliest blood islands. These cell aggregates develop into blood vessels.

A few sections posteriorly the head process will consist of the dorsal (head) ectoderm, and the dorsally approximated neural folds. This thick-walled, ectodermal tube is now the neural tube, which will be seen to bulge laterally (evaginations) to form the optic vesicles, at the level of the pros encephalon. The central cavity is the prosocoel. Immediately ventral to the neural tube there will shortly appear a group of cells which, more posterioily, surround a cavity. This is the most anterior pocket of the foregut, later to be designated as the pre-oral pit. Between the foregut and the neural tube locate the small circular aggregation of cells, the anterior tip of the notochord. Scattered cells may be found between the neural tube and the head ectoderm. They are embryonic mesoderm cells and in this region they are designated as head mesenchyme cells. The space between the head process and the underlying double-layered pro-amnion is the space of the sub-cephalic pocket.

As the germ layers of the head process are reflected into the underlying pro-amnion, more posteriorly, one will find that between the lateral ectoderm and endoderm there appear a pair of large spaces enclosed in a new double layered mesoderm. These spaces seem to come in and under the foregut region and are known as the amnio-cardiac vesicles or the primordia of the pericardial cavity. The mesoderm has grown in toward the embryo but before reaching the embryo it has formed this large cavity by splitting into an outer (somatic) and an inner (splanchnic) layer of mesoderm. These spaces are most obvious at about the level of the anterior intestinal portal, i.e. the region where the foregut is no longer a closed tube but opens out over the underlying yolk. The yolk is not present on these slides, having been removed when the blastoderm was excised. Note that the mesial, portion of the splanchnic mesoderm (i.e., that closest to the gut region) is relatively very thick. This tissue will give rise to the muscle portion of- the heart, known as myocardium. Note also that the area opaca at this. level is closer to the embryo; that the neural tube is still open dorsally; and that there is abundant mesenchyme lateral to the neural tube is still open dorsally; and that the is abundant mesenchyme lateral to the neural folds where the somites are to appear. The notochord still marks the mid-dorsal line of the arch enteric roof and the Space below (between the layer of endoderm and the yolk) is later to be included in the foregut, with the posterior migration of the anterior intestinal portal.

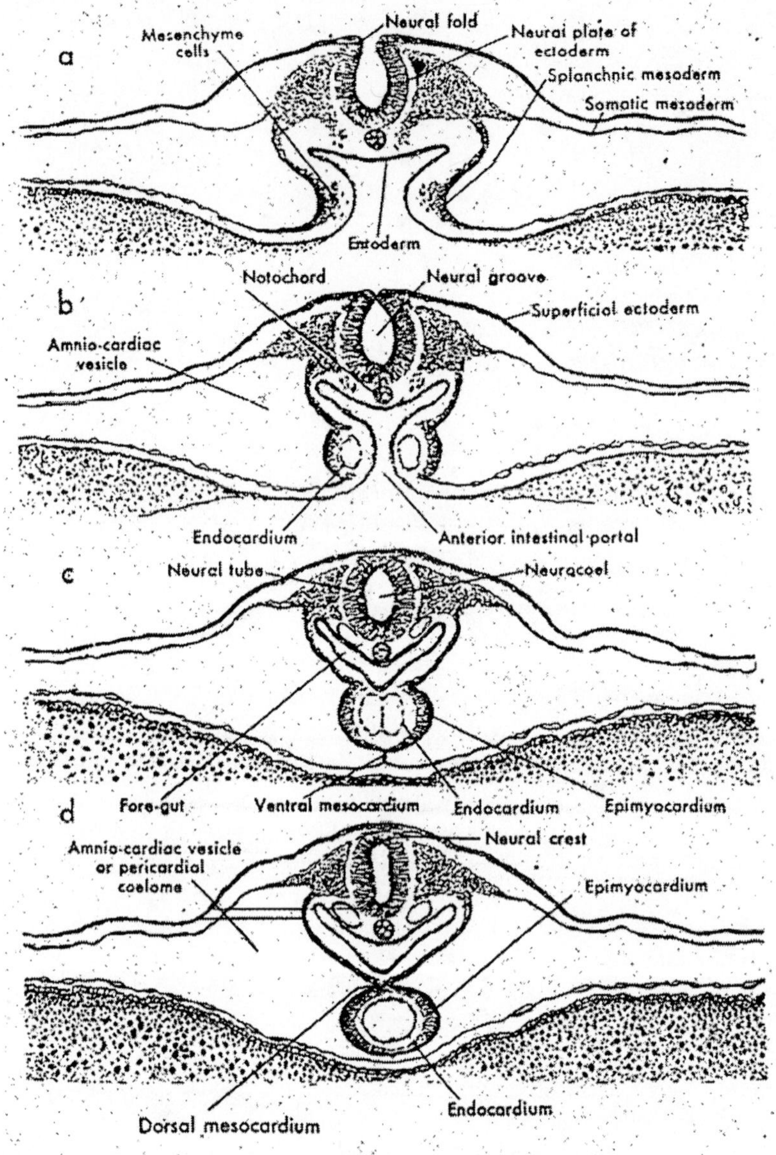

Fig. 29.2 a, b, c, d

The amino-cardiac vesicles are large and will be seen through many posterior sections, but the arch enteric roof opens out flat over the yolk so that there seems to be no indication of any gut structure except a roof. Other changes to note are the open neural tube; the early differentiation of mesoderm into the dorsal (epimeric) or segmental mesoderm (the somites), the intermediate (mesomeric)* or nephric mesoderm, and the split lateral (hypomeric) mesoderm this most lateral mesoderm is split into the dorsal somatic and ventral splanchnic layers, with the intervening coelomic space. The term coelom is properly reserved for the mesodermal cavity within the body of the embryo. Since most of the above mentioned cavities will remain outside of the embryonic body they are called the exocoels, or extra-embryonic body cavities. The line of demarcation shifts with age, the coelom and exocoel remaining continuous for many hours. Locate blood islands in the splanchnic layer only, in the area opaca.

Examine sections further posteriorly until it becomes difficult to differentiate between the notochord and' the neural tube, even under high power magnification. As the notochord disappears one approaches the position of the receding Hensen's node and the primitive streak. The neural folds will flatten out ancl merge into the primitive folds, in succeeding sections. Instead of formed somites there will appear large masses of mesoderm which seem to be merged with (growing out from) the ectoderm cells of the primitive streak. No notochord will be seen, but beneath will be the endodermal roof of the primitivegut. A section through Hensen's node will show it as an. aggregation of cells, posterior to which is an angled depression, the primitive pit. At this level also identify the area pellucida; the area opaca where the yolk granules are abundant; and the germ wall which' consists of the inner margin of the zone of junction about to be added to the area pellucida.

Transverse Sections of 33 Hours Chick Embryo (Fig. 29.3 a, b, c, d, e and f)

Begin the study at the most anterior of the serial transverse sections of the 33 hours chick embryo and locate the opticoel with the connecting optic stalks, all associated with the diocoel (cavity of the diencephalon). Dorsal to the opticoel you will find abundant head mesenchyme. The sub-cephalic pocket and the pro-amnion are much the same as they were in the 24 hour stage but laterally the exocoels have grown in toward the head process so that there is a clearer demarcation between the sornatopleure and the splanchnopleure. The former is made up of ectoderm and somatic mesoderm and the latter is composed of splanchnic mesoderm and endo-derm. Laterally the area opaca is identified by the adherent underlying yolk, at the region of the germ wall. The splanchnic mesoderm of the area vasculosa is now supplied with abundant blood islands, some with contained corpuscles.

Several sections posteriorly the opticoels will appear to be separated from the diocoel, indicating a slightly posterior direction to the optic evaginations from the forebrain. Just as the opticoels disappear you will find a pair of small blood vessels close to the mesocoel, the paired anterior cardinal veins. Directly ventral to these veins and much larger are the dorsal aortic roots, the anterior projections of which become the carotids. At about this level a median ventral invagination appears as the foregut comes into view. The fused ectoderm and endoderm of this region comprises the oral plate the invagination being the ectodermal stomodeum. As the pharynx widens you will find either a pair of ventral aortic roots or some connections between the dorsal and the ventral aortae, which represent the rudiments of the aortic arches. Attempt to identify these aortic arches by number counting from the most anterior. On some slides the amniotic folds will have closed over the embryo dorsal to the head process, indicating the beginning of the true double-layered amnion and chorion. It must be emphasized that while the layer next to the embryo is always ectoderm, the second layer of the amnion (the outer one) is mesoderm. The chorionic layers are oppositely arranged. Posterior to the first appearance of the notochord, which is between the mesocoel and the pharynx, locate the ventral aorta as a single large thick walled vessel ventral to the pharynx. This will become the bulbus arteriosus.

Following the bulbus posteriorly locate the thin inner lining of the heart (endocardium) and the thicker muscular part (myocardium). The thick muscular portion immediately associated with the hulbus is the embryonic ventricle. The heart mass is suspended to the floor of the gut by a ticable layered membrane known as the dorsal' mesocardium. The coelom in this vicinity will become the pericardial cavity, derived from the original amino cardiac vesicles. The enclosing mesodermal membrane is the pericardium. Within the splanchnopleure it will be possible to identify the vitelline blood vessels, but it may not be possible to distinguish between veins or arteries. A few sections posteriorly however, the omphalomesenteric veins will be seen to converge from their more lateral positions to enter the atrium by way of the sinus venosus. Some of the epimeric mesoderm at these levels may be organized into somites.

Posterior to the heart, the gut is open over the yolk, the margin (anterior intestinal portal), having moved somewhat posteriorly since the preceding stage. The floorless gut at this level will, in all probability, later become the mid-gut. The omphalomesenteric veins now appear to be far out in the splanchnopleure, lateral to the limiting sulcus which is the line of (future) demarcation between the embryonic and extra-embryonic structures. In some sections it will be possible to discern this partial constriction which separates the coelom from the exocoel or extra embryonic body cavity (e.e.b.c.). The dorsal aortae are very large. The small paired posterior cardinal veins may be seen just dorsal to the mesomeric (nephric) mesoderm. Well-formed somites should be apparent and above these, in the angle between the body ectoderm and the neural tube, identify the neural crests (in the same position as in the frog).

More posteriorly, as the neural groove opens and the somites disappear in favour of solid mesodermal areas, note that the notochordal tissue is displaced by a larger mass of compactly arranged cells representing Hensen's node. The lateral mesoderm here seems to be continuous with the Hensen's node material. The groove dorsally his flattened and disappeared entirely. This is about all that remains of the original primitive streak.

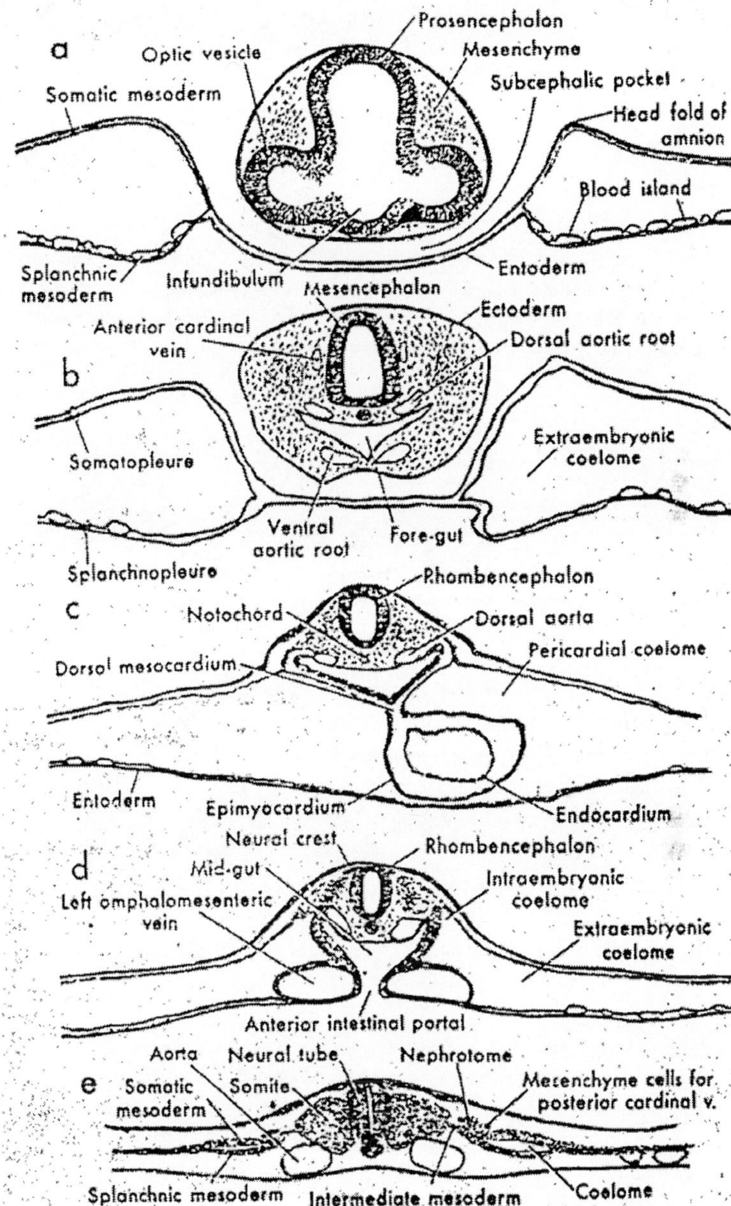

Fig. 29.3 a, b, c, d, & e

Transverse Section of 48 Hours Chick Embryo

Fig. 29.4a: A section through the anterior end of a 48-hour chick embryo. As a result of the flexures of the brain, this section passes through the diencephalon, mesencephalon, and metencephalon, and possibly myelencephalon. The torsion of the embryo places the left side of the head region in contact with the yolk, and this side may be identified in sections by the presence of blood vessels between the splanchnic mesoderm and the endoderm. The head region is enclosed in a sac of ectoderm and mesoderm, the amnion. The outer limiting membrane is the chorion, which extends over the surface of the yolk. The mesenchyme between the brain and the ectoderm contains the anterior cardinal veins.

Fig. 29.4b: A section through the anterior pituitary (Rathke's pouch) of a 48, hour chick embryo. The pharyngeal pouches (visceral pouches) extend to the ectoderm and communicate with the exterior through visceral clefts. A visceral arch of mesoderm is thus cut of, and the first aortic arch develops. This arch connects the dorsal aortae with the ventral aortae. The parts of the brain are identified as diencephalon, by the proximity of the anterior pituitary, and as myelencephalon, by its thin, dorsal wall. The anterior cardinal veins are closet() the ventro lateral wall of. the myelencephalon. The notochord is ventral to the latter. The student is reminded that, because.of the curvature of the brain, the right half of succeeding sections is more and more anterior, while the left half is more and more posterior.

Fig. 29.4c: A section through the eye of a 48-hour chick embryo. The optic vesicle changes from a bulge of the brain to a cup-shaped optic cup and remains connected to the brain by means of the optic stalk. The inner layer (nearest the lens) is the retinal layer, which develops the sensory elements of the eye and the optic nerve. The outer layer becomes pigmented. In this two pairs of visceral pouches are present at this time and they divide the mesoderm into visceral arches. Two pairs of aortic arches are seen together with the paired dorsal aortae and the paired anterior cardinal veins, which are present on both sides of the telencephalon and on both sides of the myelencephalon.

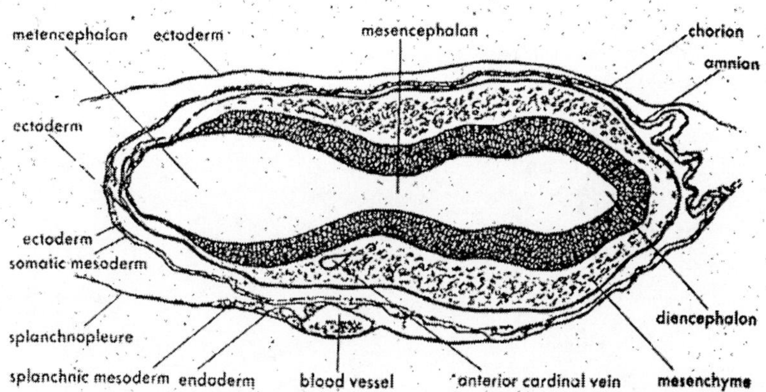

29.4 a

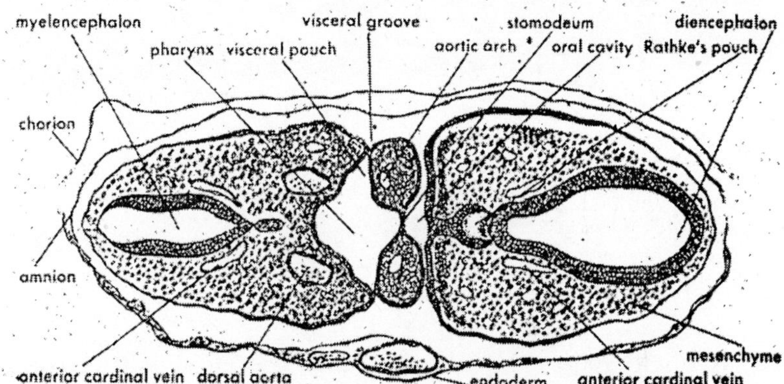

29.4b

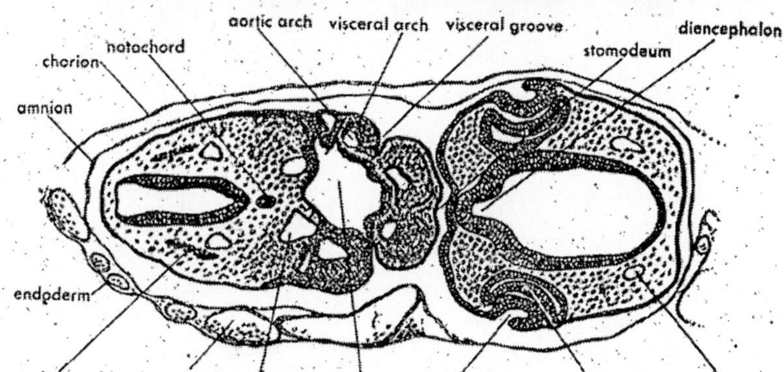

29.4c

Fig. 29.4d: A section through the otocyst of a 48-hour chick embryo. (A detail of the formation of the otocyst appears in the next figure, 29.4.e). The depression in the floor of the pharynx is the thyroid, and its later development into a tube. The optic cup in the present section lacks a lens.

Fig. 29.4e: A section through the invaginating auditory placode, of a 48-hour chick embryo. The infolding of the placode to form a vesicle is similar to that of the lens ectoderm to form a lens vesicle. The thin-walled brain is the myelencephalon, which receives fibers from a part of the auditory placode which contributes some of the cells for the seventh and eighth cranial nerves. The primordium of the nerve ganglia develops into the acoustic ganglion of cranial nerve VIII and the geniculate ganglion of cranial nerve VII.

Fig. 29.4f: A section through the thyroid and bulbus arteriosus of a 48-hour chick embryo. The section shows the connection between the dorsal and ventral aortae by way of the aortic arch. The early formation of the heart and its relation to the pharynx. The beginning of the thyroid gland is shown as a depression in the floor of the pharynx just above the heart. The thyroid separates from the endoderm as a tube and later becomes a. solid structure. This is the last section through the brain. The very tip of the telencephalon and also the beginning of the spinal cord shows in this figure.

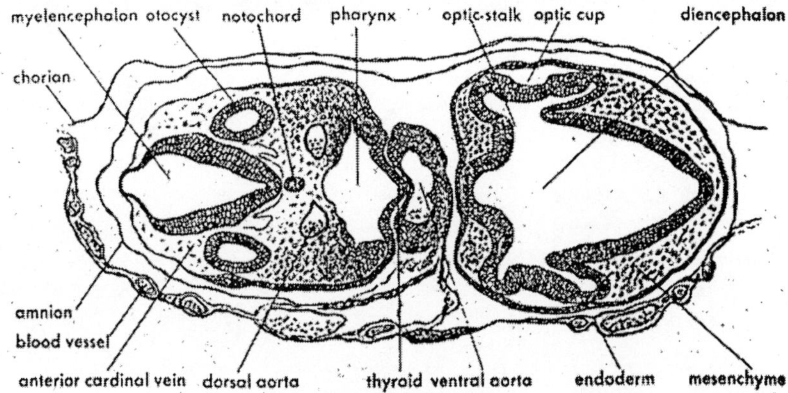

29.4d

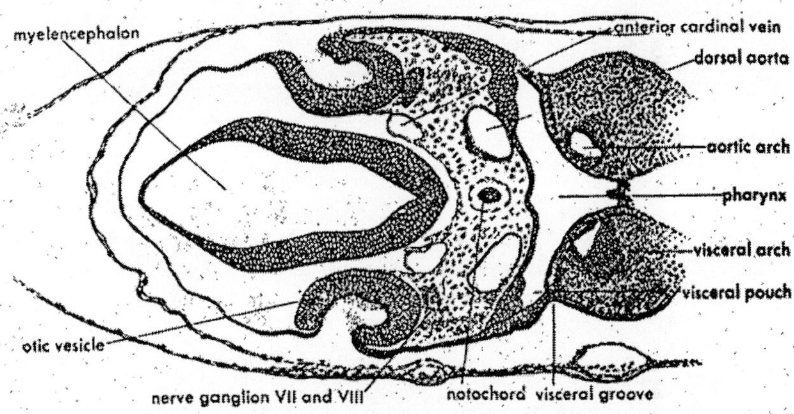

29.4 e

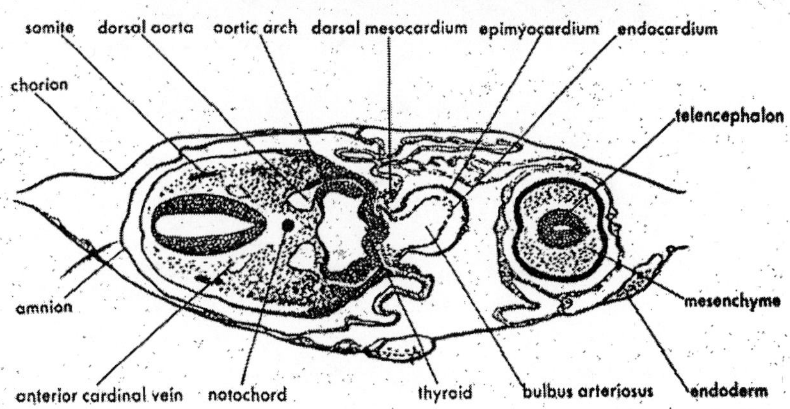

29.4f

Fig. 29.4g: A section through the heart and laryngotracheal groove of a 48-hour chick embryo. Because of the S-shaped bend of the heart, this section passes through the atrium and the ventricle. The left anterior cardinal vein and the duct of Cuvier can be seen to be approaching the heart, and in figure (29.4e) the duct joins the sinus venosus. In the present section the dorsal aortae have fused. The gut shows a ventral groove which is the primordium of the endothelial lining of the lungs and trachea. This groove will split into a bilobed structure and grow into the mesenchyme. The gut is surrounded by mesenchyme and the walls of the coelom. The coelom extends up around the gut and a dorsal mesentery is formed by fusion of the mesoderm. The somite has separated into a compartment derniatome and myotome into scattered, diffuse sclerotome.

Fig. 29.4h: A section through the duct of Cuvier and the sinus venosus of a 48-hour chick embryo. The duct of Cuvier can be seen entering the sinus on the left side. The sinus venosus also receives the right and left vitelline veins. The coelom is closing over and around the gut. This results in the formation of mesenteries by fusion of the mesoderm. The splanchnic mesoderm thus surrounds the endoderm of the gut and contributes the connective tissue, muscle, and peritoneum. The endoderm forms only the lining of the digestive tract and its glands.

Fig. 29.4i: A section through the developing liver diverticulum of a 48-hour chick embryo. The liver originates as two outgrowths of the endoderm adjacent to the vitelline veins. When the two vitelline veins unite to form the ductus venosus, the liver continues to differentiate around this blood vessel. finally, % rhen the liver becomes functional, the vitelline veins break their direct connection with the venosus, and an hepatic portal system is formed. The liver retains its connection to the gut as the ductus choledochus.

At this point the student may trace the circulation of the blood forward from the vitelline veins into the. sintis venosus, where the ducts of Cuvier enter and thence forward and in a reverse curve through the atrium and ventricle, and finally forward through the bulbus arteriosus and ventral aorta and on up through the aortic arch into the dorsal aortae.

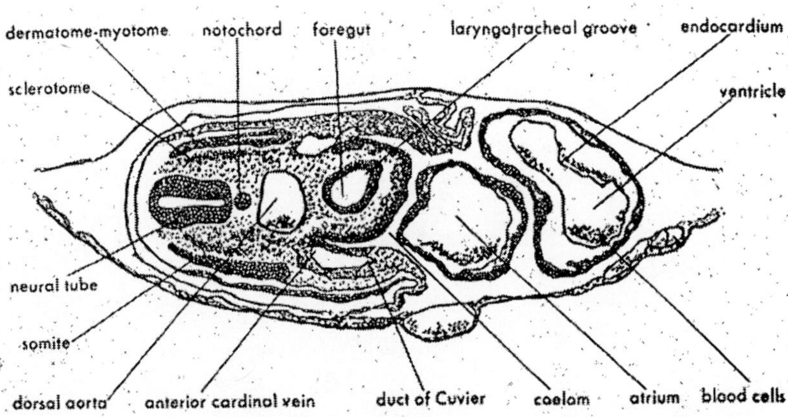

29.4 g

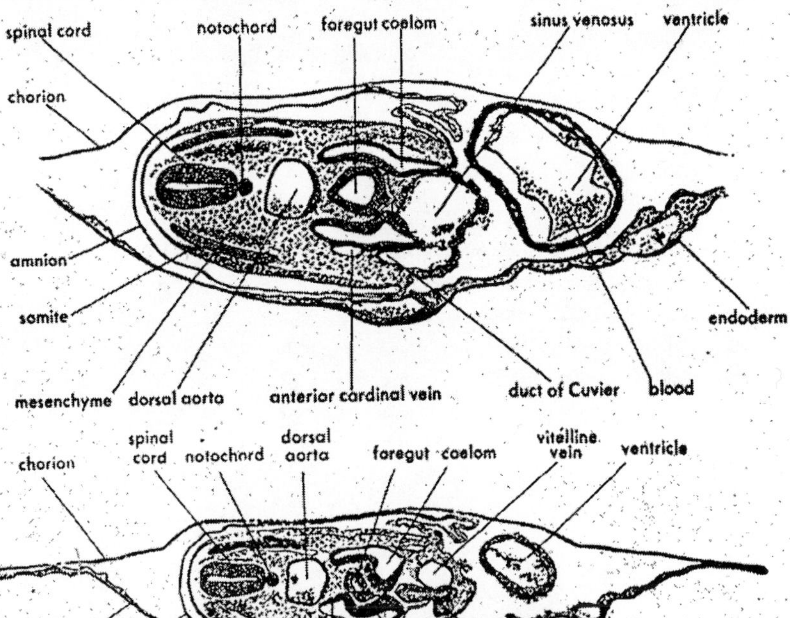

29.4h

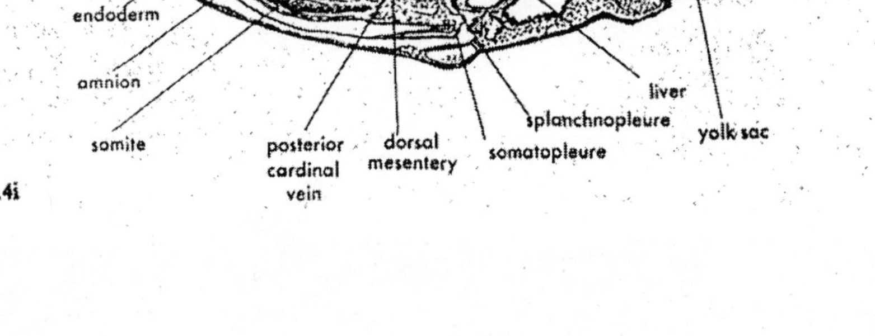

29.4i

Fig. 29.4j: A section through the seroamniotic connection of a 48-hour chick embryo. In the mesoderm, the somites, nephrotome, and lateral mesoderm have separated and have begun differentiation. The split in the lateral mesoderm results in a splanchnic and a somatic layer. The somatic layer becomes thin and is closely applied to the ectoderm. This double layer, known as somatopleure, folds up over the embryo and fuses in the mid-line, resulting in the amnion and chorion. The splanchnic layer forms a close union with the endoderm, and this double layer is the splanchnopleure. The splanchnopleure contains the blood vessels and forms the yolk sac and allantois. Within the embryo the somatic and splanchnic mesiderm form the lining of the coelom. The splanchnic layer fuses in the mid line to form the dorsal mesentery which suspends the gut. The somatic layer forms part of the body wall and contributes to the mesoderm of the limbs.

Fig.29.4k: A section through the lateral folds of the amnion of a 48-hour cluck embryo. These three stages illustrate the steps in the formation of the amnion and chorion by lateral folds. In the present section the early somite has segregated from the nephrotorne and has undergone partial differentiation into dermatome, myotome, and sclerotorne. This differentiation continues, where the three divisions are distinct. In this figure the nephrotome has undergone some differentiation into a mesonephric duct, and the mesonephric tubules are forming. Note the proximity in this figure of the posterior cardinal vein and dorsal aorta to the developing mesonephros. When functioning, blood from the dorsal aorta passes through the mesonephros and flows into the renal portal system derived in part from the posterior cardinal vein. Excretions pass posteriorly through the mesonephric duct to the allantois.

Fig. 29.4l: A section through, the vitelline artery of a 48-hour chick embryo. Differentiation at this level is considerably less than in the preceding section. The somites are more primitive and the nephrotome is not so well developed. The amnion is absent, but there are suggestions of the amniotic folds. The general appearance is more like that of an anterior section of a 33-hour embryo than of the preceding sections, thus illustrating the general law that posterior regions of an embryo develop later than anterior regions. By contrast, note a posterior section in the 72-hour embryo, where the amniotic folds are closed and the mesoderm is well differentiated into somite, mesonephros, and lateral mesoderm.

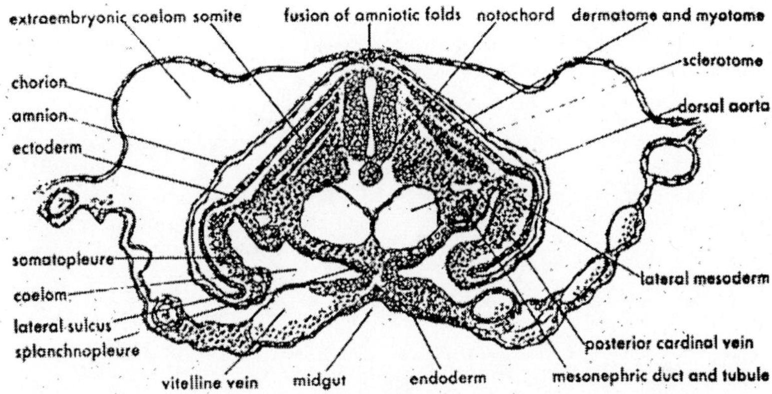

29.4j

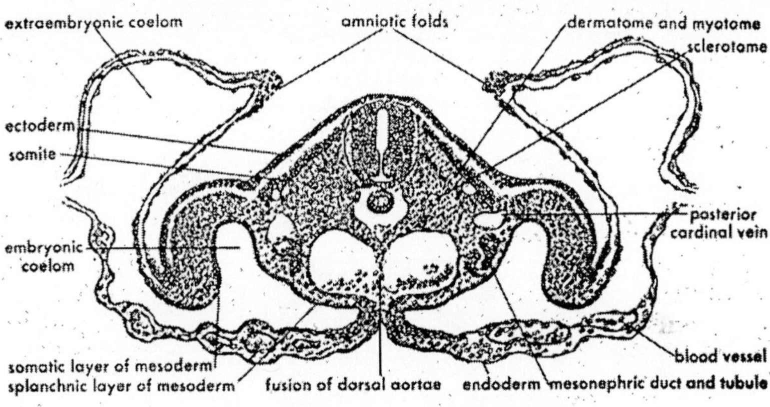

29.4k

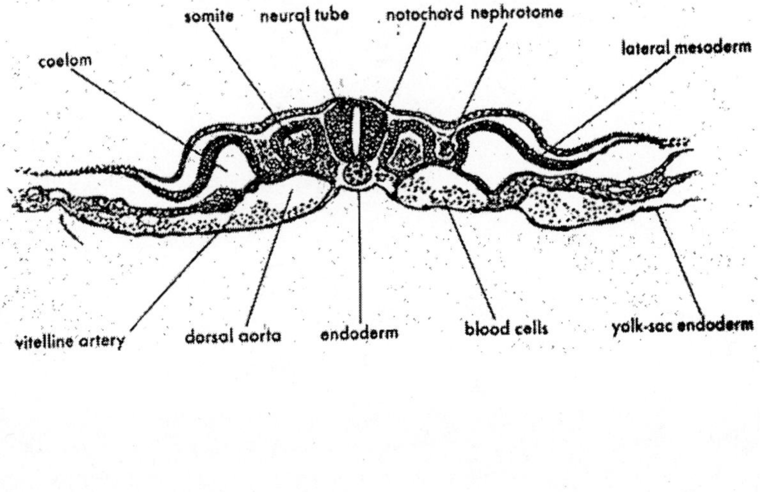

29.4l

Fig. 29.4m: A section through the tail bud of a 48-hour chick embryo. A tail fold of the sOmatopleute that is similar to the head fold of an earlier embryo results in the formation of the amnion. A tail fold of the splanchnopleure under the embryo results in a hindgut and a posterior intestinal portal, both of which resemble the anterior region of a 7 somite embryo. The tail bud is a mass of undifferentiated tissue, in which ectoderm, mesoderm, and endoderm are indistinguishable. -It is similar, therefore, to the primitive streak of the 33-hour embryo. As the hindgut closes and the posterior intestinal portal moves forward, an outgrowth of the hindgut forms the allantois. This important structure functions during embryonic development as an organ of respiration and excretion.

Fig. 29An: An intermediate stage between 48 hours and 72 hours of development. The chief advances of. the 72-hour embryo over the 48-hour embryo are increased differentiation of the brain, with sharper cranial and cervical flexures; further differentiation of the eye and the appearance of the olfactory pits; an increase in the number of somites and the appearance of the LoMb buds; further growth of thewamnion to the extent that it almost encloses the embryo, anhough an oval-shaped window in the posterior region still rediains; and further folding of the heart until the ventricle lies posterior to the atrium.

Section of 72 Hours Chick Embryo

Fig. 29.5a: A section through the auditory vesicle of a 72-hour chick embryo. Due to the cranial and cervical flexures the section is a frontal one- i.e. it is parallel to the long axis of the brain rather than across the axis. The cells anterior to the vesicle will form the seventh and eighth cranial ganglia, and the, latter will form the auditory nerve. Another cluster of deep staining cells in about the middle of the section will form the fifth cranial ganglia, and posterior to the vesicle lies the primordium of the ninth cranial nerve.

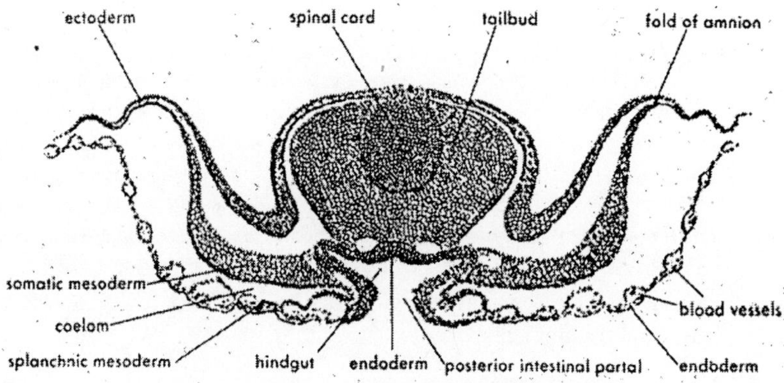

29.4m

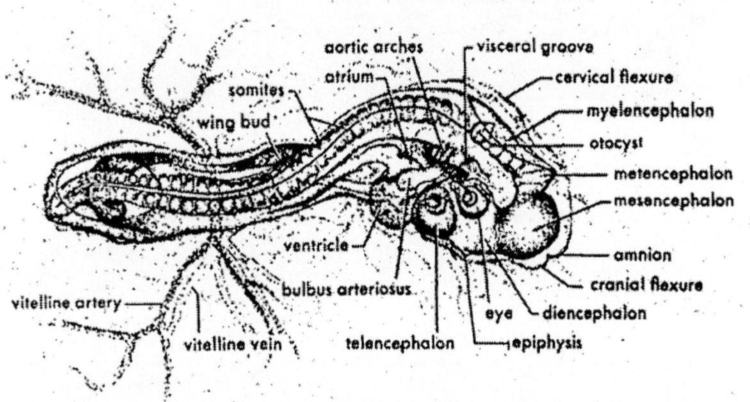

29.4n

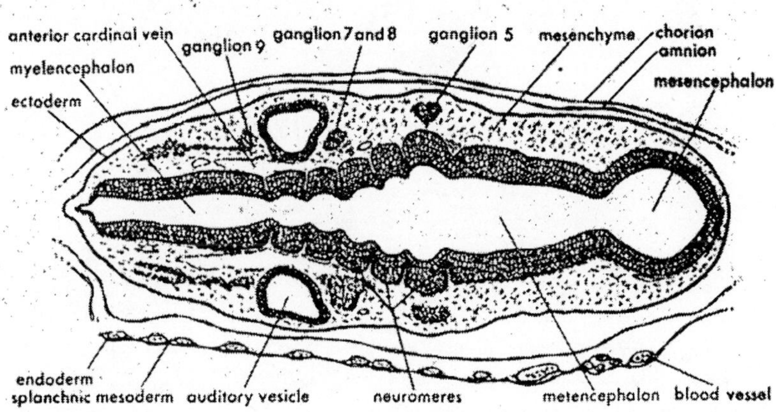

29.5a

Fig. 29.5b: A section through the pharynx of a 72-hour chick embryo. This section is at about the level at which the stomodeum enters the pharynx and at which the ectodermal pocket of Rathke's pouch communicates with the oral cavity. The next section shows the pouch itself. In all, four clear pairs of aortic arches are present at this time; the figure shows the junction of the fourth arch with the aorta. On the right side, this arch forms the main systemic arch from the heart to the dorsal aorta. Four pairs of visceral pouches and four pairs of visceral arches the mandibular, the hyoid, the third, and the fourth- are present. The visceral pouches do not form, definite visceral clefts at the union with the visceral grooves, but some of them break through, forming apertures to the exterior.

Fig. 29.5c: Similar to figure 29.5b but a little more poste. The anterior pituitary is developing from Rathke's pouch, which finally fuses with the indibulutr, Also leading in from the oral' cavity is the stomodeum, which opens into the pharynx. Four pairs of visceral arches are shown with four pairs or aortic arches. The right aortic arch is approaching the aorta and flows into it. Three pairs of visceral pouches are present.

Fig. 29.5d: A section through the thyroid of a 72-hour chick embryo. The thyroid originates as a groove in the floor of the gut of a 48-hour embryo near the second pair of pharyngeal pouches. This groove becomes a tube in the 72-hour embryo and finally separates off from the pharynx completely. In figure note the third and fourth visceral pouches, the three visceral grooves, or furrows, and the dienephalon; the region near the oral cavity is the infundibulum, which is in contact with Rathke's pouch (anterior pituitary primordium). When these two elements fuse and subsequently differentiate into a secreting gland, the pituitary is formed. Since the anterior lobe of the pituitary is derived mostly from ectoderm of the oral cavity, whereas the more posterior lobes develop from the brain, it is not surprising that the secretions of these regibns differ greatly in the adult.

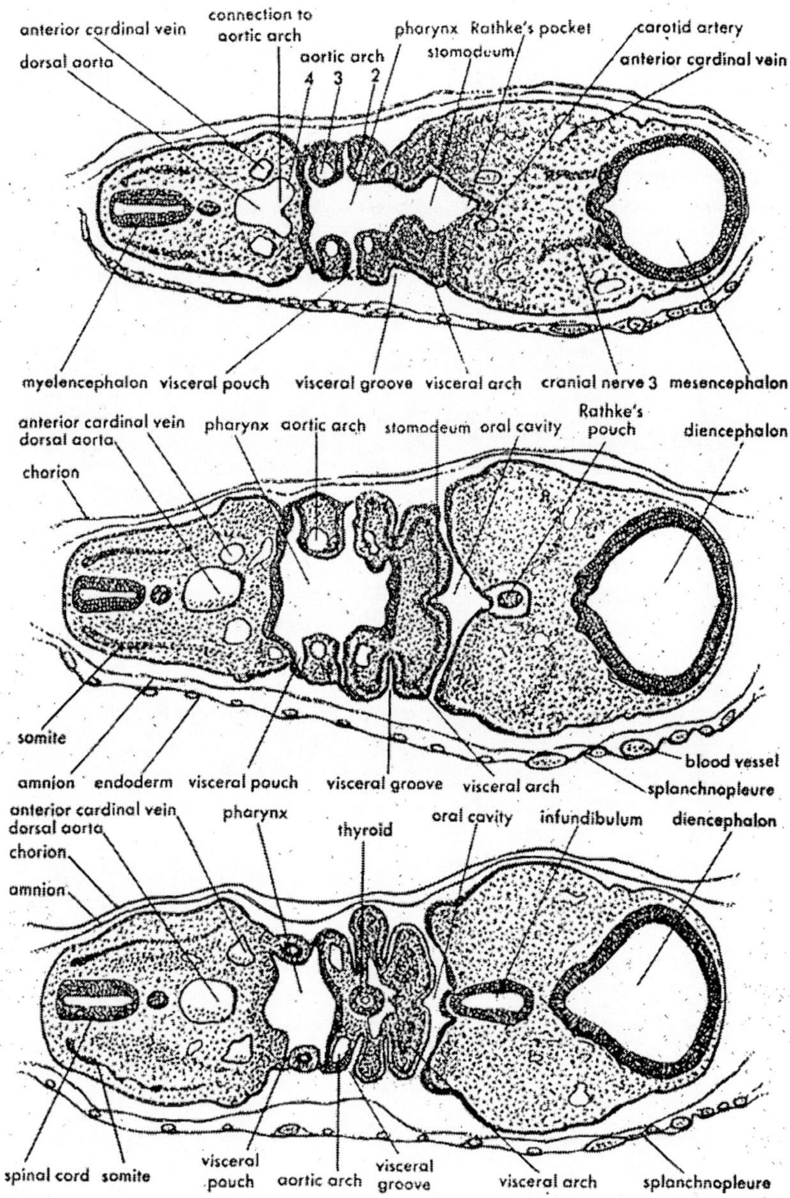

29.5b

29.5c

29.5d

Fig. 29.5e: A section through the eye, the heart, and the laryngotracheal groove of a 72-hour chick embryo. The optic cup has two layers; a lateral layer which will develop into the sensory layer of the retina and which will send nerve fibers back to the brain, and a medial layer which becomes pigmented. The optic cup arose by an indentation of the optic vesicle. The cup remains attached to the brain by means of an optic stalk, which has a cavity continuous with the cavities of the optic cup and of the brain. It is along and within this stalk that the nerve fibers grow to join the brain. The lens has separated from the ectoderm and is hollow. Its earlier origin is shown as an invagination of the ectoderm. The lens is induced to form in the ectoderm by the stimulus of the optic vesicle. The groove in the floor of the gut is part of the primordium which forms the lungs and trachea.

Fig. 29.5f: A section through the lung diverticula and the duct of Cuvier of a 72-hour chick ' embryo. The anterior cardinal veins may be traced from the region of the myelencephalon, posteriorly alongside the dorsal aorta, then below the dorsal aorta, until they finally enter the ducts of Cuvier. The blood flows into the atrium. The laryngotracheal groove becomes led, and in the later stage the two lung elements appear separated from the esophagus above. The figure shows the formation of mesenteries above the below the gut. Note that the coelom and splanchnic mesoderm curve up over the gut and its mesenchyme to form a support-the dorsal mesentery. Below, the heart is pinched off by the fusion of splanchnic mesoderm to form a ventral mesentery. The mesenchyme in this region will form the supporting tissues of the lungs, and the endoderm will form the lining.

Fig. 29.5g: A section through the lung buds and the duct of Cuvier, sinus venosus, and atrium of a 72-hour chick embryo. This section illustrates the two lung diverticula embedded in mesenchyme. On the right side the duct of Cuvier is seen entering the sinus venous, which in turn opens into the atrium of the heart The blood then flows through the ventricle then into the bulbus arterious and finally through the trumcus arteriosus into the aortic arches and up into the dorsal aorta. The anterior and posterior cardinal veins bring the blood from the head and body back to the ducts of Cuvier, which enter the sinus venosus. Note the two thickened regions of the ectoderm in the head region opposite the telencephalon. These are the nasal placodes, which invaginate to form the nasal epithelium, the olfactory nerve, and accessory Structures.

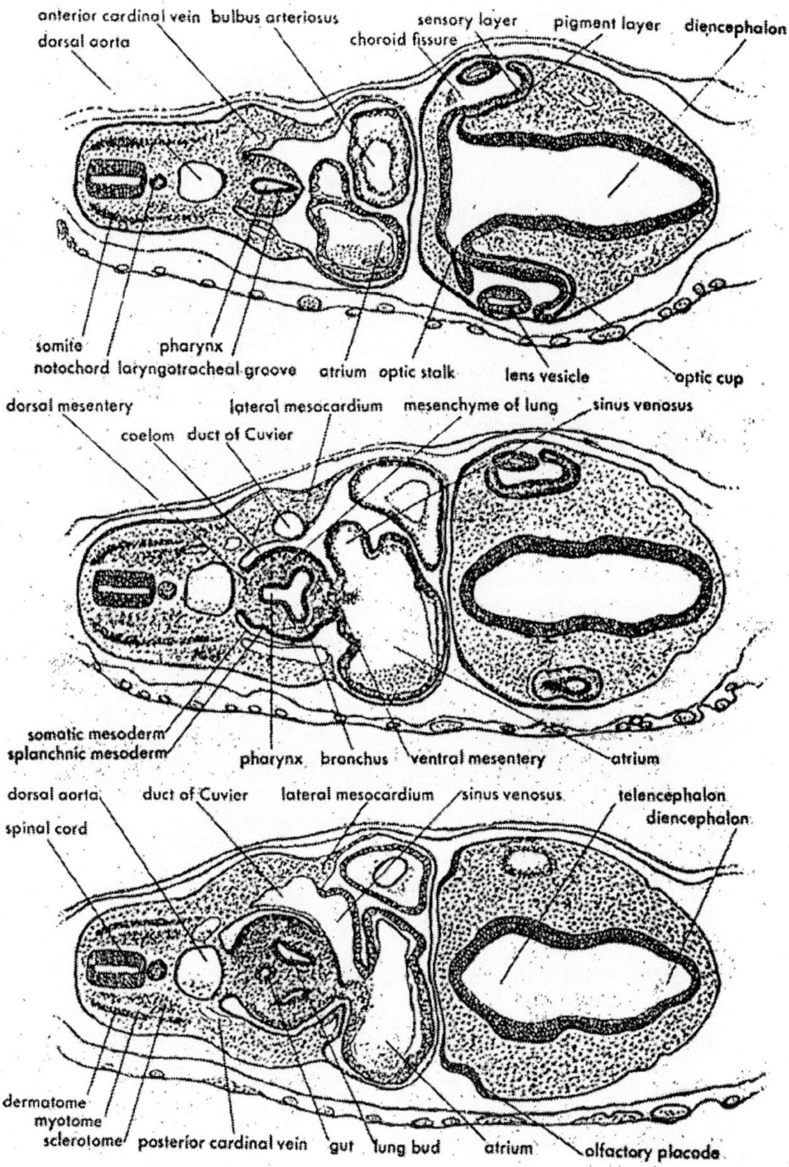

29.5e

29.5f

29.5g

Fig. 29.5h: A section through the olfactory pits of a 72-hour chick embryo. The lung diverticula are no longer present, and the duct of Cuvier is shown opening into the sinus venosus. The posterior cardinal veins are on the same level as the aorta. Dorsal and ventral mesenteries suspend the esophagus and mesenchyme which will develop into lung. The heart is cut through the two arms of the ventricular loop. This section is shown chiefly for the development of the olfactory pits. They begin as thickened ectoderm which invaginates in a fashion similar to that of the lens and the otocyst. Unlike the invagination of the lens and the otocyst, however, the nasal invagination always remains open to the exterior as the external nares and invaginates until it meets the pharynx, forming a connection, the internal nares, the olfactory nerve develops from the cells of the placode and grows back into the telencephalon.

Fig. 29.5i: A section through the liver of a 72-hour chick embryo. The liver begins as outgrowths of the gut adjacent to the vitelline veins. After the fusion of these veins to form the ductus venosus, which flows into the sinus venosus, the liver diverticulae grow around the ductus venosus and retain a single connection- the bile duct-with the intestine. Later, an hepatic portal system develops, in which blood flows through sinusoids in the liver and is collected by hepatic veins. A bile bladder also develops. The present section passes through the telencephalon, the diencephalon, and the epiphysis. The large, thick-wailed vesicle is a section through the ventricle of the heart. From here the blood can be traced forward to the aortic arches to the atrium, sinus venosus, and ducts of Cuvier to the cardinal veins.

Fig. 29.5j: Another section through the developing liver of a 72-hour chick embryo, showing the junction of a liver diverticulum with the duodenum of the gut. The section is close to the opening of the gut through the anterior intestinal portal and the fusion of the vitelline veins. The ventral folds of the amnion have not yet fused on the left side of the embryo; in succeeding sections the endoderm of the gut passes through this opening and finally joins the extraembryonic endoderm, the yolk sac. The midgut is thus continuous with the yolk sac; with further closure of the amniotic folds, the gut endoderm will be constricted to form a yolk stalk opening into the yolk sac. In the present section note the formation of the umbilical vein in the lateral body mesoderm. This vein drains the allantois.

In succeeding sections the slices are posterior to the heart and pass through the body of the embryo. These sections are nearly true cross sections at right angles to the longitudinal axis, whereas many of the preceding sections have been either frontal or oblique.

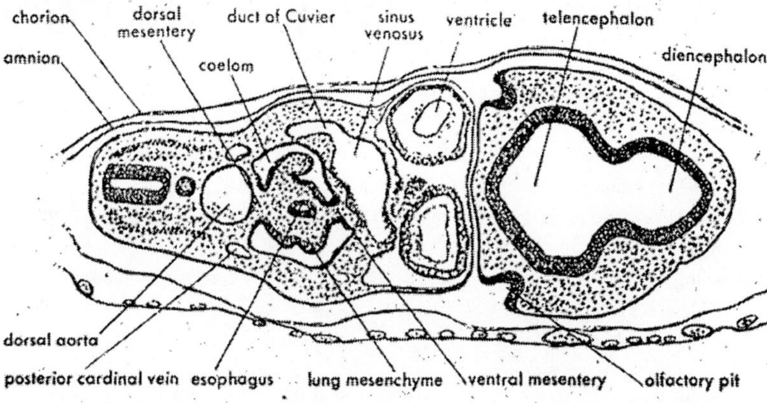

29.5h

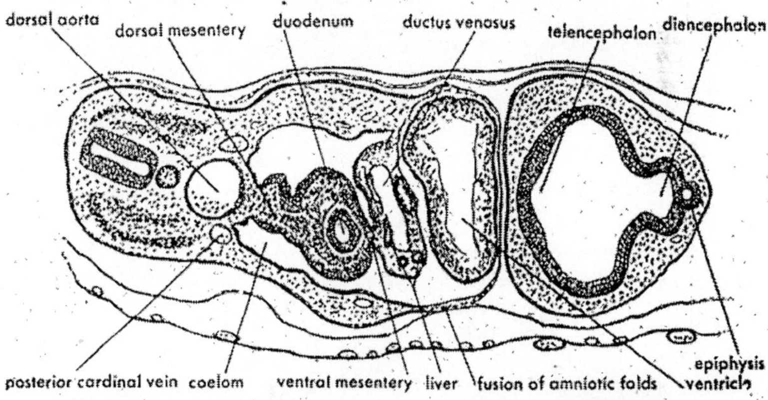

29.5I

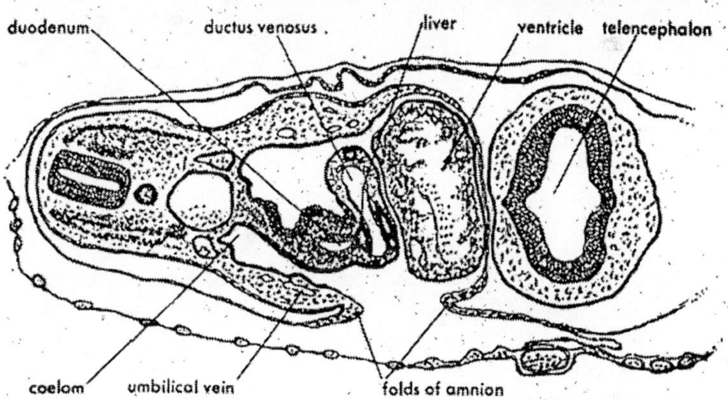

29.5j

Fig. 29.5k: The developing limb bud of a 72-hour chick embryo. The limb develops fror mesenchyme derived from the somatic layer of mesoderm. The mesenchyme pushes into th ectoderm to form a protuberance which, with continued growth, becomes club-shaped and the forms digits. The nerves supplying the limb come from the ventral neuroblasts in the spirt cod and from the neural crests. Note the distance and terrain through which a nerve fiber mu grow to reach the developing limb. The thick dorsal mesentery supports the two large vitellir veins, which run forward and fuse to form the ductus venosus. At the same time the endoderi of the midgut is shaped into a tube, and a ventral mesentery forms. The ventral folds of tl amnion come together, thus walling off the coelom.

Fig. 29.5l: A section through the junction of the dorsal aortae and the vitelline arteries. T vitelline arteries form a network of vessels which, by fusion and growth, develop into lar arteries which carry blood from the dorsal aorta into the yolk sack. Figure illustrates the uni of the dorsal aortae with the vitelline arteries. The dorsal aortae then continue posteriorly a finally form the caudal artery. This figure is at a level where the body stalk forms by t lateral limiting sulci coming close together but not cutting off the gut or vitelline arteries. vitelline veins leaves the embryo at the same level as do the vitelline arteries. The large ves adjacent to the vitelline artery is possibly the vitelline vein.

Fig. 29.5m: A section through a mesodermal somite of a 72-hour chick embryo. The son begins as a solid block of mesoderm. It next develops a split, and then becomes segrega from the nephrotome by a deep constriction. It then begins to differentiate into the three in components, dermatome, myotome, and sclerotorne. The dermatome later becomes applied the ectoderm and forms the dermis. The myotomes develop into the striated muscle of the b and other regions, while the loose sclerotome forms a sheath around the neural tube notochord to form the vertebrae.

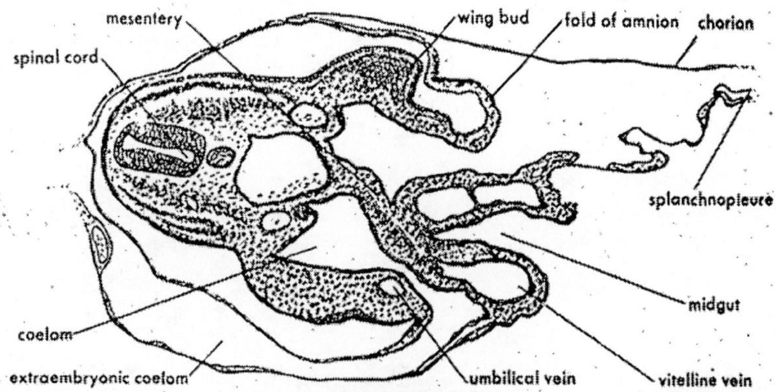

29.5k

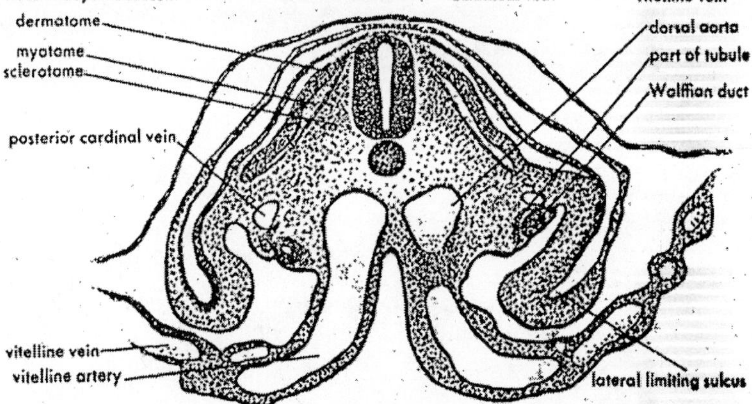

29.5l

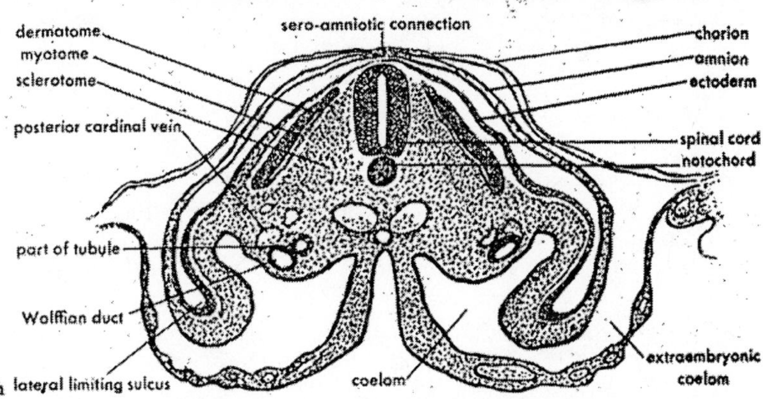

29.5m

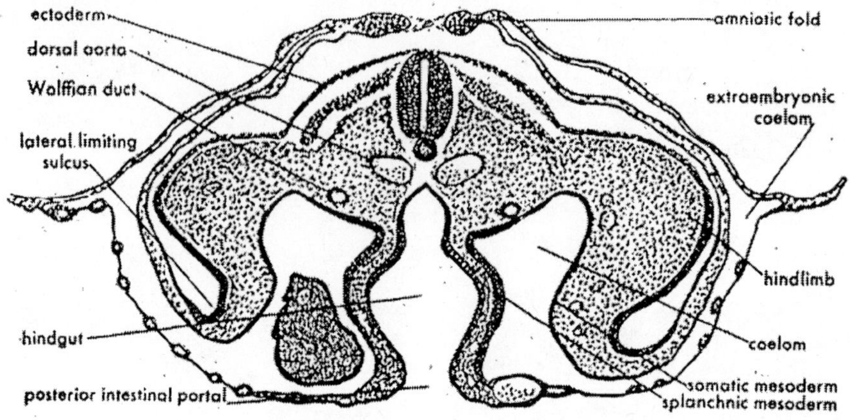

Fig. 29.5n: A section through the hindgut and the posterior intestinal portion of a 72-hour chick embryo. The action passes through the small opening between the amniotic folds into the amniotic cavity, where lateral, anterior, and posterior folds almost meet. The hind-limb buds are shown as thick masses of mesenchyme derived from somatic mesoderm. The hindgut is well formed at 72-hours by a tail fold which is similar to the head fold. Lateral folds, also aid in closing the hindgut. The hindgut receives the Wolffian ducts and later on in development the ureters and oviducts. From the floor of the hindgut, the endoderm grows out in the form of a spherical sac which becomes the lining of the allantois.

Transverse Sections of 96 Hours Chick Embryo (Fig. 29.6 a, b, c, d, e, and f)

It is important that the student have before him an accurate whole mount drawing, or a mounted specimen, while studying the serial transverse sections. The flexions and torsions are so great that practically none of the sections is truly transverse in the sense that very few of them cut across the embryonic axis at right angles. The most anterior sections will appear to be frontals. The sections at the level of the tail fold will be reminiscent of those through the earlier head fold.

The majority of the structure seen in the 96 hour stage have already been identified in earlier stages so that a mere listing of the structures, under the headings of their germ layer derivatives, of this stage will be given. Some of these structures may be difficult to locate in any particular series of sections but each student should become thoroughly acquainted with the list of derivatives presented below.

Membranes

- **Amnion:** inner layer ectodermal, outer layer mesodermal; together forming the innermost sac which encloses the embryo. Later acquires smooth muscle fibres which help to keep the embryo in more-or-less constant motion, preventing adhesions.
- **Chorion:** inner layer somatic mesoderm, outer layer ectoderm; formed

simultaneously with, and reciprocally to the amnion this membrane will appear as a thin, non-vascular layer dorsal to the embryo. Syn. serosa.

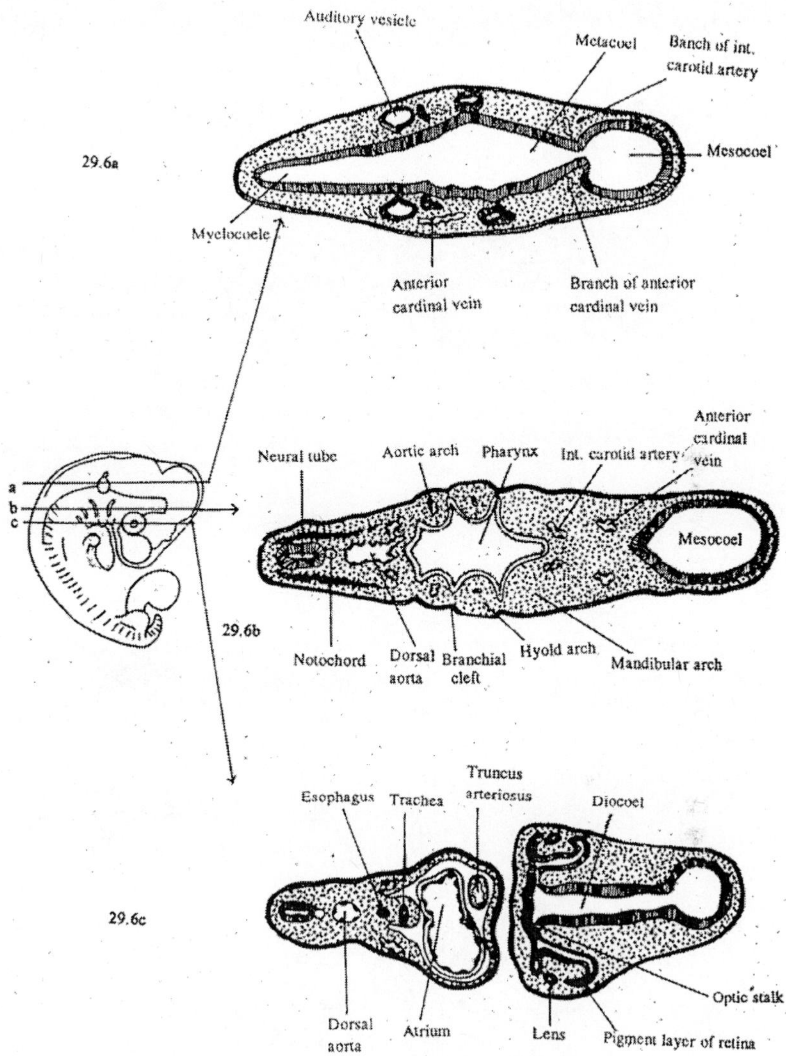

- **Yolk Sac Splanchnopleure:** dorsal splanchnic mesoderm and ventral endodermal layer comprise this membrane which, both anteriorly and posteriorly, separates the embryo from the underlying yolk. There are numerous blood vessels in the splanchnic mesoderm of this layer. Lateral to the embryo locate the junction of this splanchnopleure and the chorion, the space between being the exocoel or extra embryonic body cavity. This cavity is surrounded by somatic and splanchnic mesoderm.
- **Allantois:** this membrane is made up of a bladder-like median ventral

diverticulum of the hind-gut endoderm, covered with splanchnic mesoderm. It will be found only in sections posterior to the posterior intestinal portal at this stage of development. Ultimately this double membrane will fill the exocoel and its (outer layer of) mesoderm will fuse with mesoderm of the chorion and the amnion and finally also with the splanchnic mesoderm of tie yolk sac splanchnopleure. Its function in the chick is related to respiration and excretion.

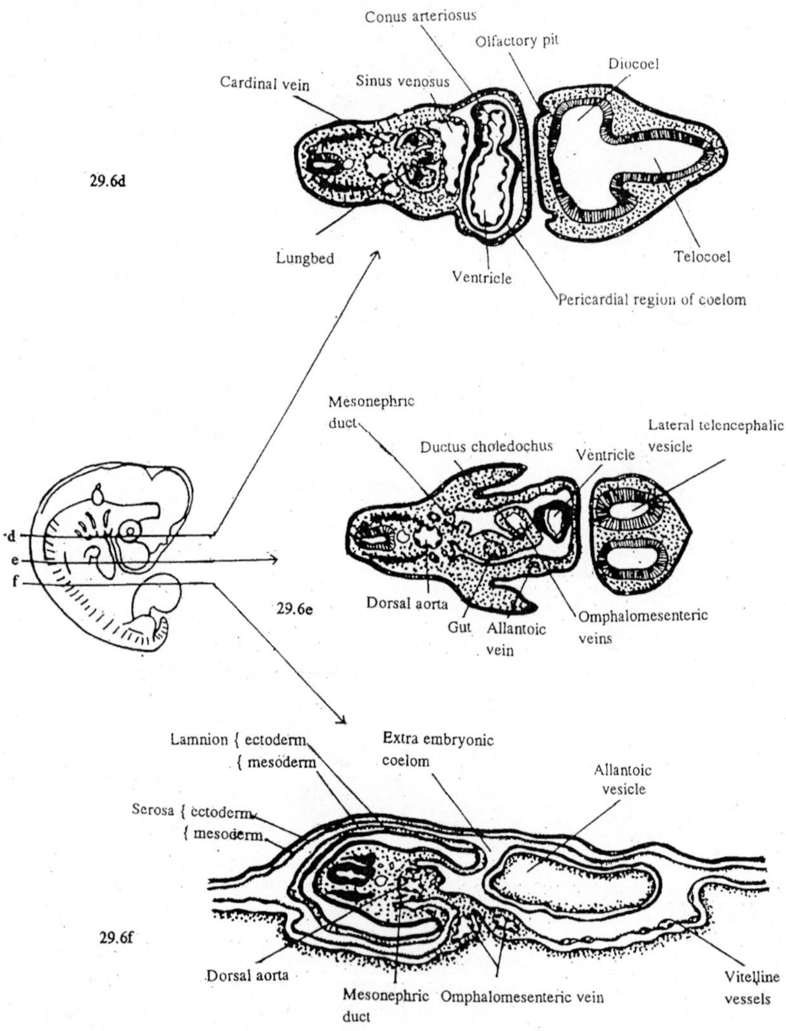

Ectodermal Derivatives

- **Epidermis:** the entire outer covering of the chick embryo is of ectodermal origin, and is made up largely of squamous epithelium but also includes horny scales, feather germs, quills and barbs, claws, beak coverings, and a temporary egg tooth.

By invaginations from the surface the linings of the following structures are also derived from ectoderm; the mouth (stomodeal portion); cloaca (proctodeal portion); visceral clefts (peripheral halves); nostrils; eye chamber and lens; and external auditory meatus.

Brain Vesicles and Related Structures Prosencephalon (Forebrain)

- **Telencephalon:** cerebral hemispheres with thick lateral walls called the corpora striata; torus transverses which becomes the anterior commissure; paraphysis which comes from an evagination of the telencepbalon median; the anterior choroid plexus anterior to the velum; olfactory lobes from the ventral evaginations associated with the first cranial nerves; telocoels which become the lateral ventricles and join the third ventricle by way of the foramina of Monro.
- **Diencephalon:** epiphysis becomes the frilled pineal; posterior commissure is the thickened roof just posterior to the epiphysis; optic thalami from thickened lateral walls; optic chiasma from optic nerve fibres; infundibulum plus Rathke's pocket forms the pituitary gland; and the diocoel becomes the posterior part of the third ventricle: (There are two telencephalic and one diencephalic choroid plexi).
- Mesencephalon (Mid-brain): the dorso lateral thickenings of this portion of the brain become the optic lobes; the ventral and lateral thickenings become the crura cerebri associated with the cranial nerves 3 and 4; the roof becomes the corpora bigemina; the cavity (mesocoel) will be the narrow aqueduct of Sylvius which joins the third and the fourth ventricles.

Rhombencephalon (hind-brain)

- **Metencephalon:** the cerebellum comes from the roof; the pons Varoli from the ventrolateral walls and the cavity is the metecoel. This portion of the hind-brain is separated from the mesencephalon (anteriorly) by the isthmus, a dorsal depression.
- **Myelencephalon:** the roof becomes the thin and vascular posterior choroid plexus; the floor thickens to become the brain stem and medulla associated with cranial nerves 5 to 12 inclusive; the cavity is the myelocoel or fourth ventricle.
- **Spinal Cord:** this mid-dorsal tubular extension of the brain eetoderm and cavity (neurocoel) with 38 pairs of spinal ganglia ad related sympathetic ganglia, extends into the tail fold. The spinal ganglia are derived from the neural crests. The sympathetic ganglia come from the same source but migrate toward the dorsal aorta.

Endodermal Derivatives

Foregut

- **Pre-Oral Pit:** Seessel's pocket, present in the embryo only.
- **Mouth cavity:** contiguous with the stomodeum.
- **Pharynx:** the dorso-ventrally flattened cavity just posterior to the mouth cavity, related to the visceral pouches and clefts.
- **Visceral Clefts:** the paired lateral evaginations of the pharyngeal cavity which meet and fuse with the invaginations of head ectoderm to form the visceral clefts. Three and sometimes four pairs present in the chick embryo.
- **Thyroid Gland:** median ventral evagination from the pharyngeal floor near the base of the second pair of visceral arches.
- **Laryngo-tracheal groove:** median ventral groove in the floor of the foregut posterior to the thyroid anlage and leading to the origin of the trachea. The connection with the gut is the glottis.
- **Oesophagus:** thick-walled oval tube anterior to origin of lung buds.
- **Primary bronchi:** the tracheal evagination grows caudal and bifurcates to form the lung buds, later associated also with the air sacs.
- **Stomach and Duodenum:** section between the origin of the lung buds and a point just anterior to the liver and pancreatic rudiments.
- **Liver:** median ventral diverticulum of foregut which appears early in development and invades the large omphalomesenteric veins causing them to become broken up into sinuses.
- **Pancreas:** three nidiments, one dorsal and the others ventral to the liver.
- **Anterior intestinal portal:** immediately posterior to the liver diverticulum the foregut opens out over the yolk, determining the anterior margin of the unformed mid-gut.

Midgut

Portion of the unformed gut bounded by the anterior and the posterior intestinal portals, therefore, merely a roof over the yolk. Has thick lateral walls but no floor. Gives rise to intestines.

Hindgut

- **Allantois:** bladder-like median ventral evagination from the floor of the hindgut, endodermal lining. This sac eventually fills the exocoel, its outer layer of mesoderm fusing with the inner lining of the exocoel.
- **Cloaca:** portion of hindgut posterior to the origin of the allantois and that region which receives the mesonephtic ducts at the dorso-lateral margins; later receives the reproductive ducts and functions also as a rectum. The common outlet for the three systems.
- **Post-Anal Gut:** extension of hindgut into tail fold, comparable to pre-oral pit.

Function unknown.

- **Notochord:** embryonic precursor of centrum of axial skeleton, composed of vacuolated cells. Precise origin doubtful.
- **Mesenchyme:** loose embryonic mesoderm, found most concentrated in the head.
- **Epimeric mesoderm:** (dorsal derivatives)
- **Dertnatome:** dorsal mesodermal mass, continuous with dermatome, precursor of striated musculature. This includes cranial, axial, and appendicular muscles.
- **Sclerotome:** loose mesenchymal cells from the somites which migrate around the nerve cord and notochord to give rise to the axial skeleton.
- **Mesonieric mesoderm:** (intermediate portion)
- **Nephrotome:** primordium of excretory system, both tubules and duos. The mesonephros may be seen at 72 hours and the metanephros at 96 hours. Portions give rise to gonoducts.
- **Hypomeric mesoderm:** (lateral plate mesoderm)
- **Somatopleure:** mesodermal sheet combined with embryonic ectoderm.
- **Splanchnopleure:** mesodermal sheet combined with embryonic endoderm, very vascular mesodermal portion gives rise to visceral musculature.
- **Coelom:** space between somatic and splanchnic mesoderm, within the body, from the walls of which develop the gonads. Subdivided into pericardial, pleural, and peritoneal cavities.
- Gonads-.primordia of sex glands from dorsal germinal epithelium (mesoderm) about the 8th day, near dorsal mesentery.

Visceral Arches (all paired)

- **Maxillary process:** mesoderm in head, will form upper jaw.
- **Mandibular process:** first visceral arch, anterior to hyomandibular pouch, will form parts of lower jaw.
- **Hyoid process:** posterior to hyomandibular pouch.
- **Third visceral arch:** (also called first bronchial) posterior to second visceral cleft.
- Fourth Visceral arch-(also called second bronchial) posterior to third visceral cleft. Circulatory System

Arteries

- **Dorsal Aortae:** paired vessels dorsal to the pharynx which fuse to form a single large vessel posteriorly. Give off the following branches.
- **Inter segmental Arteries:** paired dorso-lateral branches between somites.
- **Brachial Arteries:** branches of dorsal aorta to forelimb rudiments.

- **Mesenteric Artery:** many branches to the developing gut.
- **Vitelline Arteries:** paited vessels which leave the body at about the level of somite 22, to course out over the yolk.
- **Allantoic Arteries:** numerous small vessels from the iliac artery to the developing allantois.
- **Renal Arteries:** branches of dorsal aorta to the nephrogenic tissue.
- **Iliac Arteries:** lateral extensions of the dorsal aortae at the level of the posterior limb buds.
- **Caudal Arteries:** extensions into the tail of the fused dorsal aortae.
- **Pulmonary Arteries:** extensions of the original sixth pair of aortic arches, growing with the development of the lungs and air sacs.
- **External Carotids:** small paired vessels extending from the base of the third pair of aortic arches toward the mandibular process.
- **Internal Carotids:** anterior extensions of the paired dorsal aortae into the head. Circulatory System

Veins

- **Cardinal Veins**: paired vessels forming letter "H" from dorsal view.
- **Anterior Cardinals**: carry venous blood from the head to common cardinals.
- **Posterior Cardinals**: carry venous blood from dorsal nephrogenic tissue toward the common cardinals. Posterior portion later degenerates while the anterior portion becomes the subclavians.
- **Sub-Cardinals**: smaller veins located ventral to the nephrogenic tissue, later to fuse to form part of the posterior vena cava.
- **Common Cardinals**: fused junction of anterior and posterior cardinals which empty into the sinus venosus through the ductus Cuvieri.
- **Pulmonary Veins**: ventral to the primary bronchi, carry blood to the left auricle.
- **Vitelline veins**: very large paired vessels from the yolk sac, called the omphalomesenterics, as they enter the embryonic body. These veins converge as the meatus venosus before entering the sinus venosus. The meatus venosus later becomes the hepatic vein and those portions of the omphalomesenterics posterior to the liver and the newly formed dorsal mesenteric vein become the hepatic portals.
- **Sub-Intestinal Vein**: small vessel appearing in the ventral mesentery.
- **Dorsal Mesenteric Veins**: develops late and assumes functions of vitelline veins about the time of hatching.
- **Inter segmental Veins**: small and metameric veins which lead from the intersomitic regions to either the, anterior or the posterior cardinals.
- **Allantoic veins**: formed along with the development of the allantois, hut

reach the ductus Cuvieri by way of the lateral body walls of the embryo rather than within the body proper. Junction later changes to the hepatic vein near the heart, connections with ductus degenerating.

- **Heart**: this is a tubular organ attached to the body at the level of the bulbus arteriosus anteriorly and the sinus venosus posteriorly, the intermediate portion remaining somewhat outside of the body.
- **Sinus Venosus**: receives common cardinals, through ductus Cuvieri; also the meatus venosus which is the fused omphalomesenteric veins.
- **Atrium**: (aniage of the) auricles, ventral to the sinus venosus; thin-walled.
- **Ventricle**: projected farthest from the embryonic body; ventral to atrium; very thick-walled.
- **Bulbus Arteriosus**: (also conus) junction between ventricle and the aortic arches; circular in transverse section and identical with the ventral aorta.
- **Aortic Arches**: these vessels pass through the visceral arches from the bulbus arteriosus (ventral aorta) to the dorsal aortae, located directly dorsal to the pharynx. Since the aortic arches run through correspondingly designated visceral arches (previously identified) a total of five (from 2 to 6) may be located. The last two may be fused.

APPENDIX I

Chromatography and Electrophoresis

Chromatography

Chromatography is the most versatile technique for rapid separation of cellular substances. The separation of amino acids illustrates a typical procedure. A small portion of the solution containing the acids is placed on a strip or sheet of filter paper. An organic solvent partially miscible with water is allowed to flow up the paper by capillary action. As the solvent flows it carries the individual amino acid components along with it at various speeds according to their solubility in a solvent. The resolved amino acids may then be observed either by examining them under ultraviolet light or by spraying the paper with a chemical reagent (ninhydrin for amino acids) which reacts with the components to form a coloured compound. Chromatography involves 2 phases; a mobile phase and a stationary phase.

Circular Paper Chromatography

Requirements: Whatman No. 1 paper, Mixture of amino acids, Solvent (n-butanol acetic acid, distilled water), 4:1:5, (Prepare a solvent in a separating funnel. Remove out the lower layer, preserve upper layer). Amino acids-alanin, aspargin, glutamic acid, trytophan, 0.25% Ninhydrin in acetone, Hair Drier, 0.1 ml pipette, Cotton wick, A petridish, a square wooden case with glass lid.

Method (Circular Chromatography)

1. Cut a circular piece of Whatman filter paper and saturate with acetone and dry it.
2. Mark a point in the centre and draw small circle around the central point.
3. Mark two points on the one side of the circle.
4. Spot the mixture on one mark and individual amino acid on other mark and dry the spots with a hair drier.
5. Make a hole in the centre of the paper.
6. Now fill the petridish with the solvent and dip the cotton wick in it for about 5 minutes.
7. Place the petridish in the centre of the glass case and place about 8 to 10 small bottles around it to balance the circular paper over the petridish.
8. Now pass the wick through the centre hole of the circular chromatography paper, one end of the wick remains dipping in the solvent in the petridish.
9. Let the solvent run for about 2 hours.
10. After 2 hours mark the level of solvent spread in the circular manner.
11. Remove the paper and dry in oven.

12. Spray with ninhydrin and dry in oven.
13. Coloured spots occur showing presence of more than 1 amino acid in the mixture.
14. Find out the Rf (rate of flow) values.

$$\text{Rf} = \frac{\text{Distance travelled by substance}}{\text{Distance travelled by solvent}}$$

Paper Chromatography of Amino Acids (Ascending)

Material Required: Mixture of amino acids, solvent, spraying reagent, whatman paper no. 1, sprayer etc.

Method: Cut the whatman paper no. 1 into a convenient size and draw a line just 2 cm above from one end. Wash it with acetone, dry it and spot with amino acids given. Dry the spot and place it inside the jar and close tightly for saturation. The solvent will rise by capillary action. After it has run to $1/3^{rd}$ of the paper, remove the paper and dry it. Spray carefully with Ninhydrin.

Dry in oven for five to ten minutes at 100°C. Spots will appear. Rf value can be calculated as:

$$\text{Rf} = \frac{\text{Distance travelled by substance}}{\text{Distance travelled by solvent}}$$

Solvent is prepared by:

n-butanol	:	Acetic acid	:	Distilled water
5	:	4	:	1

Paper Chromatography of Amino Acids (Descending)

Materials Required: Mixture of amino acids, solvent (n-butanol 4: Acetic acid-5: Water-1) Whatman's paper no. -1, sprayer etc.

Method: Set up a descending paper chromatographic chamber. Make it saturated with acetone by closing tightly. Spot amino acids as before keeP it for 5 to 7 hours for proper separation of amino acids. The solvent runs in a descending manner. Dry the paper, spray Ninhydrin reagent and keep it in oven for 5 to 7 mts.

$$\text{Calculations: Rf value of amino acid} = \frac{\text{Distance travelled by amino acid}}{\text{Distance travelled by solvent}}$$

Thin Layer Chromatography (TLC)

The principles of paper and column chromatography are also applicable to thin-layer chromatography. The main advantage of TLC over other chromatographic methods are rapid separation, small sample requirement and broad specificity for quantitatively resolving molecules and macromolecules.

Requirements: TLC developing chambers, glass sheets, applicator, silica gel mixture to be plotted, capillary tubes 20-u litre.

Method

1. Apply 20 u litre of unknown extract (in this case concentrated chloroform spinach leaf extract) with a capillary tube at a point 2 cm from the bottom of the TLC sheet. The spot should be uniforth and small. Allow the spot to air dry.
2. Transfer the TLC sheet to a developing chamber containing petroleum ether, benzene, acetone (120:40:20 v/v). The chamber should be saturated with solvent vapour beforehand.
3. When the solvent front has moved near to the top of the sheet, remove the chromogram and air dry.
4. The pigments/spots may be removed from the TLC sheet (elation) by cutting out the spots and then extracting with ethanol.
5. Identification of the extracted pigments/material may be determined by spectrophotometric analysis.

Column Chromatography

In Column Chromatography a solid material is used to adsorb a dissolved substance out of solution. Adsorption is the increased concentration of a substance at the surface of the solid. The amount of adsorption depends on (1) the nature and surface area of the solid, (2) the nature and concentration of the component being adsorbed, and (3) the temperature.

Requirements: Column chromatography tubes, sintered glass filter, separating funnels, pasteur pipettes, suction flasks, fresh spinach leaves. Solid adsorbent, slurry (sephadex), solvent.

Method

1. The solid adsorbent varies depending on the type of components being separated. All solid materials must he packed uniformly into the glass columns. Cracks allow the solvent to move at the greater rate through that part of the solid, which causes uneven, inseparable layers.
2. Slurry, Sephadex, a gel available commercially in various pore sizes is used widely at present for separating proteins or amino acids. A slurry of powder and water is poured into the column and allowed to settle.
3. **Solvent:** The solvent chosen depends on the solubility of the material to be adsorbed and on the solubility and activity of the adsorbent. All solvents must be pure; hence organic solvents should be dried and redistilled solvents must always cover the surface of the adsorbent during the experiment. Without the solvent the adsorbent dries and shrinks, allowing material to run down the sides of the glass without being adsorbed.
4. **Preparation of Chlorophyll extract:** Crush 4-5 spinach leaves in a mortar.
5. Add to 45 ml of hexane, 5 ml of benzene and 15 ml of methanol.
6. Stir and allow to stand for 30 minutes, remove supernatant by decanting. If supernatant is turbid, filter or centrifuge.

7. Transfer extract to separating funnel. Add water to wash out the methanol.
8. Dry extract by pouring it into a small beaker containing ½ cm layer of anhydrous sodium sulphate. Decant supernatant. Sample is ready.
9. Use a 20 X 130 mm chromatograph column with a coarse sintered glass filter near the bottom (to retain the packing materials). Pack the tube with the aid of a heavy glass rod as follows.
10. Pour small portions of freshly dried aluminium oxide into the tube. Tap the Al_2O_3 down after each addition until a 1.5 cm depth is formed. The column must be uniformly and firmly packed.
11. Add a 3 cm layer of dried calcium carbonate in the same manner. Finally, ad a 6 cm layer of freshly dried powdered sugar and tap down. Cover the sugar layer with a small amount of glass wool pressed flat on the column surface.
12. Mount the column on a suction flask using a rubber stopper.
13. Run freshly distilled hexane (b.p. 68 to 69°C) through the column from a separating funnel, maintaining a gentle suction. So that the filtrate comes through at 1-2 drops/second positive pressure, applied with a rubber bulb may also be used if the column is running slowly. The chromatogram must be constantly covered with at least 1 cm. of solvent above the top layer of adsorbent.
14. Add the green extract to the column at a point when there is only about 1 cc of hexane above the column. Use a relatively concentrated solution to get resolution. When all the extract has been added, develop the chromatogram by percolating a 4.1 mixture of hexane and benzene through the column. After a good separation of the various coloured bands has been obtained, percolate petroleum ether alone through the column to check any extension of the zones.
15. Suck the column partially dry using a vacuum.
16. Remove the column of adsorbent from the glass tube and cut the column between the coloured zones. The column may be extruded with a thin spatula or by forcing the contents out with the glass pestle. It sometimes helps to invert the column and let it fall several times from a height of 1 to 2 cm onto a glass plate covered with cloth.
17. Dissolve the pigments from the separated bands with 9 ml of ethyl ether plus 1 ml of methanol. Centrifuge to remove the adsorbent.
18. Run an absorption spectrum of the different coloured fractions and compare with the known spectra of pigments thought to be present.

Polyacrylamide Gel Electrophoresis

Disc electrophoresis involves first concentrating protein or nucleic acid samples into very thin layers using a discontinuous voltage gradient and then electrophoresing them on a column of polyacrylamide gel. The concentration step allows electrophoresis from starting zones only 1-100 micron thick, which produces the enhanced resolution characteristic of the technique. By varying the concentration of acrylamide and methylene bisactylamide in the gels, fractionation of the proteins on the basis of their size as well as their charge, can be activated, because larger molecules will encounter greater resistance in moving through the gel network than smaller ones.

	Solutions lower gel			Solutions upper gel	
A)	1 NHCL	24.0 ml	B)	1 M H_3PO_4	25.6 ml
	Tris	18.2 gm		Tris	5.7 gm.
	TEMED	0.23 ml		TEMED	0.46 ml.
	Water to 100 ml (pH 8.9)			Water to 100 ml (pH 6.9)	
C)	Acrylamide	30.0 gm	D)	Acrylamide	10.0 gm
	Bisacrylamide	0.8 gm		Bisacrylamide	2.5 gm.
	Water to 100 ml			Water to 100 ml	
F)	(To be prepared fresh)		E)	Riboflavin	4 mg
	Ammonium persulphate			Water to 100 ml	
	0.14 gm Water to 100 ml.				
H)	50 X concentrated upper buffer:				
	Tris	30 gm			
	Glycine	144.0 gm			
	Water to 1000 ml				

When a voltage is applied across the solution, the ions move, positive ones to the negative pole and vice-versa. The movement, of the ions constitute the current. The movement of the migration of glycinated anion is lower than that of chloride under identical conditions.

Disc electrophoresis requires "three phase" system. An upper gel, in which the protein concentrate initially; a lower gel, in which the actual electrophoresis separation occurs and an electrode buffer. The following method is for a normal analytical system.

7.5% acrylamide .... lower gel pH 8.3
2.5% acrylamide .... upper gel pH 6.9
Tris-glycine buffer .... pH 8.3

a. Use 20 ml/litre final buffer (pH 8.3): make two litres for a run. (1 n HCl 83 ml concentrated HCl/litre of distilled water)

- (1 M H_3PO_4- 68.3 ml 85% H_3PO_4/litre of distilled water).
- Tracking dye: 0.001% Bromophenol blue in water.
- Staining solution: 1% amido black in 7% acetic acid.

1. For analytical columns, glass tubes 8 cm by 0.6 cm (inner diameter), marked 3

cm from one end (the top) are used Place the tube in rubber serological stoppers so that they stand upright or if the fluorescent light is available, simply wrap the bottoms of the tubes tightly in paraffin and clamp the tubes to the face of the holder, using the metal spring across the front.

2. Prepare 1.0 ml of the lower gel solutions per tube by mixing 1 part A, 1 part C, 2 parts fresh F (***note:*** add the F last). Using the disposable pipette, add the gel solution to the 3 cm. mark of the tubes, making sure that no air bubbles have been trapped in the tubes.
3. Carefully – over layer the gel with 4.6 mm water, using a pasteur-pipette. The sharp boundary line will be blue. because of diffusion, but a new sharp line will appear 10-60 minutes later about 2 mm below the original interface. This indicates that the gel has polymerised.
4. Remove the upper liquid with a tick of wrist and place the tube in or on the fluorescent light holder. Upper gel solutions consists of 1 part B, 2 parts D, 1 part E, 4 parts water 0.3 ml of the solution is pipetted into the top of the lower gel and is then over layered with 4-6 ml of solution E or water.
5. Turn on the fluorescent light and let the upper gel solution photopolymerise for 30 minutes. It will be opalescent, flick off the liquid. The gel must now be used within hrs.
6. If stoppers have been used insert a syringe needle between the stopper and tube t, prevent a vacuum forming when the stopper is pulled off. If there are air bubbles in the bottom of the tube remove with a pasteur pipette.
7. Add enough upper buffer (about 1 litre) to the lower plastic dish so that the tubes mal contact with it.
8. Replace the upper buffer chamber on the stand and add another litre of tris-glycine buffer to it. Remove air bubbles in the tops of the tubes with a pipette.
9. The samples to be electrophorised should be made in 10% sucrose. They are then layered onto the upper gels of appropriate tubes. About 50 micrograms of protein is applied per tube. Add 1.0 ml of tracking dye to the buffer in the upper buffer chamber.
10. **Attach the electrodes:** positive polarity to the lower dish negative to the upper and turn on the power supply. Run 3-5 milliamperes per tube for 1.5 to 3.0 hours until the blue tracking dye has migrated to within 2-5 mm of the bottom of the lower gel.
11. Stop the electrophoresis and measure the distance the tracking dye has migrated into the lower gel. Rinse the gels by inserting the gel and the glass-tube. Hold the needle steady and rotate the tube when withdrawing the needle also. It will probably be necessary to rinse the gwl from both ends of the tube. Do not try to use a longer needle as these are hard to contact. The needle should be attached to a slow stream of water for lubrication during the removing of gels. Alternately use a 10 gauge needle with a rounded end (gel tubes are rimmed while immersed in water). If the gel is not coming out even

after repeated rinsing, use a rubber pasteur pipette bulb filled with water to force the gel out.

12. Place the gel in the protein stain solution for one hour. The tubes may be destained by allowing to stand in several changes of 7% acetic acid for 24 hours or by electrophoresis in the constricted destaining tubes until all the free dye is removed, using 7% acetic acid in both buffer chambers.

Other Systems

While the standard system described above is most popular discs electrophoresis of nucleic acids are running pH 4.3, 7.5% gel. Stock solution as in standard, except for:

A

1 N KOH .... 48 ml
GI acetic acid .... 17.2 ml
TEMED .... 4.0 ml
Water to 100 ml (pH 4.3)

B

1 N KOH .... 48 ml
GI acetic acid .... 2.87 ml
TEMED .... 0.46 ml
Water to 100 ml (pH 6.7)

H

10 X concentrated upper buffer
Balanine .... 312 gm
Glacial acetic acid .... 8.0 ml
Water to 1000 ml (pH 5.0)
Dilute 1:10 before use

Lower gel: 1 part A, 2 part C to 1 part of water, 4 parts F

Upper gel: 1 part B, 2 parts D, 1 part E, 4 parts water

Note: Run with electrodes reversed for proteins. (-ve to lower and +ve to upper)

A

1 N KOH .... 48 ml
GI acetic acid .... 213 ml
Water .... 400 ml
(pH 2.9)

B

1 N KOH .... 48 ml
GI acetic acid .... 2.95 ml
Water .... 100 ml
(pH 5.9)

G

Ammonium persulphate	1.4 gm
Water	50 ml
Glycine	28.1 gm
Water	1000 ml
(pH 4.0)	

H

10 X concentrated upper buffer	48 ml
Glacial acetic acid	3.06 ml

Lower Gel	**Upper Gel**
4 parts A	1 part B
2 parts B	2 part D
2 parts C	1 part E

Note: *Run with electrodes reversed for proteins.*

In preparing samples for disc electrophoresis, it is best to dialyse against the tris glycine upper buffer. In any event the presence of large amounts of chloride, sulphate etc. in the sample should be avoided, as they delay stacking for periods proportional to the amount present.

The acrylamide polymerization occurs by a free radical mechanism. Consequently, the reaction goes further at higher temperatures and it is inhibited by a variety of "quenches" including oxygen, sulphydryl compounds and reducing agents in general and some metal ions.

Most commercial preparations of acrylamide contain enough impurities to retard polymerisation and some have enough to prevent it. Acrylamide may be recrystallized by heating in Benzene.

The acrylamide vapours are quite poisonous the benzene vapour is not harmless either. So do their crystallization in a hood. Methylbisacrylamide, TEMED and ammonium persulphate are also poisonous, so avoid getting these reagents on external or internal body surfaces.

Methylenebisacrylamide and glycine can be recrystallized from hot water and tris' from 95% ethanol.

Ammonium persulphate is a strong oxidizing agent and under some circumstances can produce inactivation of enzymes and even artificial protein bands. This can be checked for in either of two ways. Mix approximately 0.5 micro mol of tris thioglycolate or polymerize the lower gel with riboflavin and light. If the latter method is used; use the same concentration of riboflavin as is used in polymerizing the upper gel, an excess inhibits polymerization. Proteins electrophoresed in riboflavin polymerized lower gels have higher rates than they have in persulphate polymerized ones.

The time required for polymerization also varies with the age of the solution F so it should be made up fresh every week or so for reproducible polymerization rates. If recrystallized acrylamide is used, the amount of solution F used may have to be reduced.

The glass tubes are washed by soaking in chromic acid cleaning solution overnight, rinsed with glass distilled water, soaked about a minute in 10% KOH in 95% ethanol and rinsed thoroughly, again in glass distilled water.

Starch Gel Electrophoresis

A major development in electrophoresis was the discovery of the remarkable resolving powers of starch gel for serum proteins.

Materials Required: Shallow perspex trays or troughs with dimensions 1 X 6 cm X 6 mm. The bottom of the trough has a hole cut in it. A false bottom is provided together with three or four cutting spacers. A perspex lid is also used with the trough starch gel electrophoresis could be vertical or horizontal.

Gel, Sample Solution.

Method: Pour gel in trough and allow to set. Cut a slit in it with a razor blade and the sample is applied into the slit. Samples of solutions may be pipetted directly into the slit which is then sealed by means of paraffin wax. Alternatively, the sample can be applied into the slit either by absorption on filter paper or cellulose acetate or as a slurry in starch grains. When the sample is applied in liquid form and the current applied, the proteins migrate to one side of the liquid origin so that a concentration gradient exists across the origin. This result in protein dimenting to the bottom of the trough at the area of highest concentration. The overall result that at the end of a run the fastest running proteins tend to be located at the bottom of the gel while the slow running proteins are located at the top of the gel.

The buffers used should cover a range of pH 2-11. Some buffers used are sodium potassium phosphate buffer pH 7.8; 0.025 M sodium acetate buffer pH 5.5; 0.076 M tris buffer, 0.005 M citric acid pH 8.6. The choice of buffer in starch gel electrophoresis is determined by the nature of the protein to be separated.

Preparation of Starch Gels

12 g of starch (BDH) is added to 100 ml of appropriate buffer in a 500 ml Buchner flask attached to a water pump. The contents are vigorously stirred and when a uniform suspension is obtained, the flask is heated evenly over a Bunsen flame and vigorously stirred. At the gel point of about 65°C the suspension becomes very viscous, a stronger stirring action is required during the subsequent heating stage, this producing a more fluid gel. When the gel becomes mobile again, and the temperature reaches 90°C the heating is discontinued and a vacuum applied to the flask for a few seconds when contents of the flask boil. Then vacuum is discontinued and gel is immediately poured in to the trough with the false bottom are lightly smeared with liquid paraffin before pouring the gel. The trough is filled until the gel is just above the top. The perspex lid is coated underneath with liquid paraffin and is now lowered gently over the surface of the gel. The lid is pressed

down firmly to extrude excess starch. The gel is left for 2-3 hours at room temperature to set.

After the gel has set lid is removed from the trough and a slit about 1-2 mm thick is cut in the gel at the origin using a razor blade and taking care to leave about ½ cm clear at both sides of the trough.

The sample is then applied to the slit (i) as a solution (ii) after absorption on paper or (iii) as a slurry in starch grains. The slit is then sealed either with soft paraffin or with a thin strip of glass. The trough is then transferred to the electrophoresis tank in either the horizontal or vertical position and contact is effected between the gel and buffer compartment by means of a double thickness of Whatman No. 3 mm paper soaked in buffer. The lid is then replaced and the run is started. Satisfactory separation with gel 20 cm long is achieved by applying a voltage of 100 volts (some 5 mA per tray) for a period of 17-19 hours when using pH 8.65 buffer. Serum samples can be stained with bromophenol blue in the usual way prior to application so that the course of separation can be followed visually.

Staining Starch Gels

After separation is complete the gel is sliced before staining using a gel slicer. For protein. Amido black, for lipoproteins oil red. For specific enzymes, standard histochemical stains.

APPENDIX II

Solutions

General Information

Solutions play a large role in many scientific investigations; therefore, it is necessary for the scientist (or prospective scientist) to become familiar with what solutions are and with some of the techniques used in preparing them.

In this laboratory block you will use several different solutions, some of them in varying concentrations. The term concentration means the amount of one substance that is contained in a definite amount of another substance. There are numerous ways of expressing concentration, for example;

Percent strength: The number of parts by weight of one substance that is contained in 100 parts by weight of solution.

Parts (grams, micrograms, milliliters, etc.) per liter: The weight or volume of one substance contained in one litre of solution.

Molarity: The number of moles of one substance contained in one litre of solution. Mole means the molecular weight of a substance expressed in grams. Litre is a unit of volume in the metric system. It is the volume occupied by one cubic decimeter of a substance. Your teacher will show you a container which measures this volume.

The terms solute and solvent are used to designate the components of a solution. The solute is the substance that is dissolved. The solvent is the liquid in which the solute is dissolved. A solution is the product obtained when a solute is dissolved in a solvent. The following conversion factors will aid you in understanding the discussion of dilution which is used to obtain weights of chemicals that are too small to be weighed on the balances in your laboratory:

1 gram = 1,000 milligrams = 1,000,000 micrograms

1 milligram = 1,000 micrograms

1 litre = 1,000 milliliters

Preparation of Solutions of Lime Concentrations

In your work you will be using extremely small quantities of some of the chemicals, many of these below the weighing capacities of the balances at your disposal. Therefore it will be necessary for you to start with a volume of solution containing an amount of the chemical which you can weigh. Assume that you have prepared 1 liter of a uniform solution that contains 1 gram of the chemical you are using. Since 1,000 milliliters of this solution contains 1,000 milligrams of the chemical, you may conclude that 1 milliliter contains 1 milligram. Can you see how volume measurements may be used to obtain the content of small quantities of a substance?

Even smaller weights' of the chemical can be obtained easily by preparing more dilute solutions. For example, assume that we place 1 milliliter of the above solution, thus 1 milligram of the chemical in 1 liter of water, we then have 1,000 milliliters of solution containing 1,000 micrograms of the chemical. We can now obtain 1 micrograms of the chemical by using 1 milliliter of this solution.

A stock solution is a relatively concentrated solution that is used in preparing larger volumes of relatively dilute solution.

Sterilizing Solutions

In biological work it is often necessary to use sterile solutions. This is particularly essential in solutions which are to be injected into living tissue. Two common methods of sterilizing solutions are autoclaving and pasteurizing.

Autoclaving: This is a method of sterilizing objects by steam under pressure. Prepare the solution and place it in the container in which it is to be maintained. Place the cap or stopper on the container loosely so that steam can escape, and place the container of solution in an autoclave or pressure cooker. Bring the temperature of the autoclave up to 120°C and maintain it for 20 minutes. If a pressure cooker is used, a pressure of 15 pounds will produce approximately this temperature. After 20 minutes remove the heat source and allow the apparatus to cool. When the pressure gauge returns to 5 pounds, you may open the steam valve and allow the remainder of the steam to escape. Allowing the pressure to go down suddenly may cause the caps or stoeps to blow off; so note carefully the above instructions. Remove the solution from the autoclave or pressure cooker and tighten the caps or stoppers immediately.

Pasteurizing: This technique is especially useful for solutions that are to be injected. Prepare the solution and place it in a sterile serum bottle or vial. Cap the bottle with a sterile rubber cap of appropriate size. (You can sterilize the bottle and cool satisfactorily by immersing them in 70% alcohol). If the rubber cap is new, prick a fine hole in it with a sterile needle, so that the expanding air can escape when the vial is heated. Place the sealed container of solution in a little water on a heat source, and bring the temperature to about 60°C. Maintain approximately this temperature for 30 minutes. Then remove the container from the water, cool, it and place it in the refrigerator or deep freeze until needed.

Preparation of Solutions Needed for this Laboratory Block

Thyroid Hormones for Amphibian Metamorphosis

- **Thyroxin Stock Solution:** Dissolve 0.1g of thyroxin in 10 ml of 0.1 M Na OH and combine this with 90 ml of water. Combine 1 ml of this solution with 1 liter of water to prepare the final Stock Solution which contains 1 microgram of thyroxin per milliliter.

- **50 micrograms per liter culture solution:** Combine 50 ml of thyroxin stock solution with 950 ml of pond water.
- **25 micrograms per liter culture solution:** Combine 25 ml of thyroxin stock solution with 975 ml of pond water.
- **10 micrograms per liter culture solution:** Combine 10 ml of thyroxin stock solution with 990 ml of pond water.
- **Triiodothyronine Stock Solution:** Substitute triiodothyronine for thyroxin and follow directions given for preparing thyroxin stock solution.
- **25 micrograms per liter culture solution:** Combine' 25 ml of triiodothyronine stock solution with 975 ml of pond water.
- **10 micrograms per liter culture solution:** Combine 10 ml of triiodothyronine stock solution with 990 ml of pond water.
- **2.5 micrograms per liter culture solution:** Combine 2.5 ml of triiodothyronine stock solution with 1 liter of pond water.
- **Iodine Stack Solution:** Dissolve 0.1 gram of iodine crystals in 10 ml of alcohol and combine this with 90 ml of water. Combine 1 ml of this solution with 1 liter of water to prepare the final stock solution which contains 1 microgram of iodine per ml of solution.
- **2 micrograms per liter culture solution:** Combine 2 ml of iodine stock solution with 1 liter of pond water.
- **1 microgram per liter culture solution:** Combine 1 ml of iodine stock solution with 1 liter of pond water.
- **Thiourea Culture Solution:** Prepare a 3.3 gram per liter culture solution by dissolving 3.3 grams of thiourea in 1 liter of pond water,
- **Control Solution:** Use pond water to which nothing is added. Holtfreter's Solution
- **Standard Stock Solution:** Combine 3.5 g NaCl, 0.05 g KCl, 0.1 g $CaCl_2$ and 0.02 g $NaHCO_3$ with 1 liter of distilled or deionized water.
- **10% Holtfreter's Solution:** Prepare this solution by combining 1 part standard stock solution with 9 parts distilled water.

Narcotic Solution

- **100 mg per liter Chloretone solution:** Dissolve 100 mg of chloretone (Chlorobutanol or 1, 1, 1-Trichloro-2-methyl-2-propanol) in 1 liter of pond water or 10% Holtfreter's solution.
- **50 mg per liter Chloretone solution:** Combine equal volumes of the above solution and pond water or 10% Holtfreter's solution.

Respiratory Inhibitor

- **Sodium Azide (NaN_3) Stock Solution:** dissolve 780 mg NaN_3 in 1 liter of acidified saline, prepared by combining 2 parts 0.9% saline with 1 part of the

Phosphate Buffer Solution C. This will give you a 1 X 10-2 M sodium azide solution.

- **1 X 10^{-3} M Sodium azide solution:** Combine 100 ml of stock solution with 900 ml of acidified saline.
- **5 X 10^{-4} M Sodium azide solution:** Combine 50 ml of stock solution with 950 ml of acidified saline.
- **1 X 10^{-4} M Sodium azide solution:** Combine 10 ml of stock solution with 990 ml of acidified saline.

Phosphate Buffer Solutions

- **Solution A:** Dissolve 9 g of Na_2HPO_4 in 1 liter of distilled or deionized water.
- **Solution B:** Dissolve 9.07 g of KH_2PO_4 in 1 liter of distilled or deionized water.
- **Solution C:** Prepare a solution at pH 5.8 by combining 7.8 parts of solution A with 92.2 parts of Solution B.

Sex Hormones and Controls

- **Testosterone propionate injection solution:** Dissolve 250 mg of testosterone propionate in 10 ml of sesame oil. Place this solution in a serum vial and sterilize it by pasteurizing as on page 76. Inject chickens with 0.1 ml or 0.2 ml of the solution as directed in the Daily Instructions.
- **Control solution for testosterone propionate:** Place 10 ml of sesame oil in a serum vial and pasteurize it. Inject control chickens with 0.1 ml or 0.2 ml of this solution, daily.
- **Chorionic gonadotrophin injection solution:** Chorionic gonadotrophin will be supplied in a sealed vial containing 500 Cortland-Nelson units (10,000 international units). Using a sterile hypodermic syringe, and needle, inject 10 ml of sterilized 0.9% saline through the rubber inset in the cap. Inject chicks with 0.1 or 0.2 ml of this solution as directed in the Daily Instructions.
- **Control solution for chorionic gonadotrophin:** Place 10 ml of 0.9% saline solution in a serum vial and pasteurize it. Inject control chickens with 0.1 ml of this solution, daily.

Thiourea for Chicken Embryos

- Thiourea Injection Solution. Dissolve 1.90 grams of thiourea in 100 ml of 0.9% saline and place it in a serum vial. This is a 0.25 M solution. Sterilize the solution by one of the methods described on pages 75 and 76.
- **Control Solution for Thiourea:** Place several ml of 0.9% saline in another serum vial and sterilize as before.

Cytochrome Oxidase Indicator

- **Alpha-naphthol Stock Solution:** Dissolve 1.44 g of alpha-naphthol in 1 liter of physiological saline. (Heat slightly to bring about solution).

- **PPD Stock Solution:** Dissolve 1.36 g of para phenylenediamine -HCl in 1 liter of physiological saline. This solution must be prepared fresh for this work.
- Nadi Indicator. Solution. Combine equal portions of Alpha-naphthol, Stock Solution, PPD. Stock Solution and Phosphate Buffer Solution C, on this page. (***Note:*** This solution must be prepared just prior to using as it darkens gradually with age).

Physiological Saline

- **0.9% Saline Solution:** Dissolve 9 grams of sodium chloride in 1 liter of distilled water. Culture Solutions for Hydra
- **Standard I Solution:** To 1 liter of distilled or deionized water add separately 1 ml each of Solutions A and B. Do not combine. Solutions A and B before placing them in the water because some of the components will precipitate out
- **Solution A:** Dissolve 33.3 g $CaCl_2$, 7.5 g KCl and 0.8 g of KI in 1 liter of distilled water.
- **Solution B:** Dissolve 25 g NaH_2PO_4. H_2O, 112 g Na_2HPO_4 and '2 g of Disodium Versonate in 1 liter of distilled water.

Buffers for Hydra

- **Buffer Solution A:** Dissolve 710 mg of Na_2HPO_4 in 1 liter of Standard I Solution.
- **Buffer Solution B:** Dissolve 680 mg of K_2HPO_4 in 1 liter of Standard I Solution.
- **Buffer Solution C:** Dissolve 121.14 g of Tris (hydroxymethyl) aminomethane in 1 liter of Standard I Solution.
- **Buffer Solution D:** Use 1 N Hydrochloric acid which is prepared by, slowly pouring 100 ml of concentrated HCI into 900 ml of Standard I Solution. (***Note:*** Always pour acid into water never the reverse).
- **Culture Solution with a pH of 6.2:** Combine 18.5 parts of Buffer Solution A with 81.5 parts of Solution B.
- **Culture Solution with a pH of 7.0:** Combine 9.4 parts of Buffer Solution D with 10 parts of Buffer Solution C.
- **Culture Solution with a pH of 8.3:** Combine 4.1 parts of Buffer Solution D with 10 parts of Buffer Solution C.

Techniques for Preparation and Use of Special Equipment the Use of the Temperature Gradient Box

This box is designed for use inside a refrigerator to provide several different constant temperatures. There are two adjustments which you can make in order to obtain a desired gradient of temperatures. First, the height of the blocks under the box may be changed to provide the proper amount of cold air needed to cool the bottom containers; and second, the temperature adjustment screw and the size of the light bulb in the top of the box may be changed to control the amount of heat needed to warm the containers in the upper levels of the box. The temperature inside the refrigerator also will have a bearing on the temperatures you are able to maintain.

Though the material in this block, for which you will use the box, requires a gradient of temperatures, the box can be used for other purposes. By covering the box with cardboard and using it outside the refrigerator, you can maintain constant temperatures equal to, or higher than, room temperature. With, a few modifications, you can use the box for incubating small numbers of hen eggs.

You should set the temperature gradient box up in the refrigerator and check it for several days before you intend to use it. For the best results with the box, the inside temperature of the refrigerator should be between 0° and 8°C, but satisfactory results may be obtained with temperatures up to 10° or 12°C.

Use an extension cord to plug the box into a 110-120 volt electrical outlet; then place two inch blocks under each corner of the box. Set stacks of culture dishes, half filled with water, in each corner. Place a thermometer in the top of the box and check the temperature at which the light shuts off. Turn the temperature adjustment screw clockwise to decrease the temperature of counter-clockwise to increase it. The temperature of the air at the top of the box should be within two or three degrees of the temperature at which the water in Dish 8 is to be maintained. As soon as the air temperature at the top of the box approaches the maximum temperature of the water you wish to study, wait an hour or so and check the temperature of the water in each dish. Once the temperature of the top dish is about right, change the height of the blocks, if necessary, to get the temperature you desire at the bottom. Check the temperature of the water in the dishes every day to see if the temperatures are remaining constant.

Improved Incubator for Hen Eggs from an Old Refrigerator

- **Step 1.** Drill two holes (½-3/4 inches) near the bottom and two holes near the top of an old refrigerator. Fit a piece of rubber tubing through one hole in the bottom of the refrigerator and insert a heavy extension cord through it. Plug the extension cord into a 110-120 volt electrical outlet.
- **Step 2.** Plug a broader thermostat into the extension cord in the box and plug a fan type hair dryer into it. Place an open container of water in the bottom of the refrigerator to maintain the necessary humidity.
- **Step 3.** Prepare egg trays as illustrated in figure IV-2. To make these trays, make a frame of 1 X 3 inch lumber and staple on a piece of 1" mesh chicken wire. Cover the edges with 1 X ½ inch strips of wood.

  Make enough of these trays to handle the number of eggs you will need for all your classes doing the block. (***Note:*** Eggs incubated in these trays will have to be turned by hand each day).

- **Step 4.** Operate the incubator for several days before placing eggs in it and adjust the thermostat so that the temperature throughout the box is maintained, at 37°-39°C.
- **Step 5.** Place the eggs on their sides in the trays, and stack the trays in the incubator. Eggs should be turned daily, at least during the first 4 days of incubation.

Any other sufficiently large, well-insulated box or container, which has a large enough opening, will replace the refrigerator for this purpose.

Alternate Method for Creating a Gradient of Temperatures (for study of heartbeat) Prepare a 40-watt fluorescent tube box, or similar-sized box, as illustrated in figure IV-3.

- **Step 1.** Wrap aluminium foil around a piece of cardboard which will just fit inside the box.
- **Step 2.** Attach a light fixture inside one end of the box (a gooseneck lamp will do).
- **Step 3.** Space and cut holes in one side of the box to hold culture dishes securely.
- **Step 4.** Half fill 6 culture dishes with 32°C physiological saline (0.9% NaCI) and place them in the box as illustrated; then turn on the light.
- **Step 5.** Open hen eggs which have been incubated 72 hours, as illustrated in figure 1-4 and place one egg in each dish of saline. Empty the contents from the shell into the saline by rotating the shell (carefully, to prevent breaking the yolk).
- **Step 6.** Allow 10-20 minutes for a gradient of temperatures to develop; then record the temperature of the saline in each dish. Count the number of heart beats per minute of the chicks in each dish and record these above the appropriate temperature on the graph.

Check and record temperature and heartbeats again later in the period if possible.

Preparation and Use of Stained Agar Slides for Staining Living Tissue

- **Step 1.** Slowly add 4 g of agar to 100 ml of boiling, distilled water, stirring constantly until the last flake of agar is dissolved. Remove the agar from the heat source and allow it to cool slightly; then pour a thin sheet of the agar over a number of microscope slides. Set these slides aside to dry for two or three days.
- **Step 2.** As soon as the slides are thoroughly dry, place them in a 1.0% solution of Nile blue sulfate or neutral red stain and allow them to remain one or more days.
- **Step 3.** Remove the slides from the stain; then rinse them off with distilled

water and set them aside to dry. Once the slides are dry, they can be placed in covered boxes and stored for later use.

Using Stained Agar to Stain a Chick Embryo

Place a drop of water on the agar and allow several minutes for the agar to soften. Use a scalpel or razor blade to cut around a piece of agar the size of the tissue you wish to stain. Scrape this piece of agar off the slide, pick it up on the point of your scalpel, and place it on the exposed embryo. (Embryo should not be covered with fluid). Leave agar sheet on embryo for about 3 minutes, then pick it off with forceps. This method is useful for visualizing very young chick embryos that are almost transparent (i.e., less than 40 hours old).

Knife for Cutting Planaria

A handy knife for cutting planaria or other small organisms can be made by breaking a double-edged razor blade in two, and then breaking off pieces just large enough to be held in the fingers. The broken piece can be mounted in a piece of cork about ½ inch thick. Alternatively the bro a piece can be glued onto the end of a glass rod; use sealing wax or epoxy resin or some other hard adhesive for the mounting job.

APPENDIX III

Necessary Equipments

The major aim of practical exercises is to train the students into self-learning and investigative procedures, independent of "book knowledge", that is, in methods designed to gain knowledge firsthand the exercises also develop confidence of students in scientific methods.

The students must, therefore, learn from the very beginning from the choice and procurement/collection of the experimental material up to the conclusions to be drawn from the experiments. He should be guided into literature search. He must be able to design experiments independently. It is most vital to faithfully record the data in a manner intelligible not only to the student after a gap of few years but also to others. Interpretation of results should be such that no conclusions are drawn that are not supported by the data at hand; yet all meaningful deductions should be arrived at A reference list helps integration of experimental results in existing body of knowledge.

It is important that instruments are properly selected and well maintained. In developmental biology many small instruments are to be fashioned by the student himself. No good work can result without proper instruments, howsoever dextrous one may be.

The following instruments are normally required

Forceps

Watchmaker's forceps No. 5, available at surgical supply stores, are good. These have very fine tips. The forceps is useless the tips meet exactly at the ends on gentle as well as strong pressure. The tips should be protected in a rubber sleeve (made from cycle valve tube) and frequently examined under microscopes. The tips can be maintained by regrinding on sand paper or an oilstone; when sharpening, keep the tip points closed. The finger and thumb should not tire if the forceps are held with tips joined for 10 minutes.

Knives

Surgical knives with blades No. 11 and detachable handle are good for general purpose. These should be meticulously maintained and blade replaced whenever necessary. Cataract knives are also good.

Scissors

Ophthalmic spring scissors and de Wicker's scissors are very good for fine dissections. Scissors with long handles and 2 cm long cutting blades are good for general dissection. Large scissors with cutting blades of 4 cm length are useful for general purpose. All scissors must cut at the very tip. The tips must be protected.

Needles

Fine sewing needles mounted on wooden handles are good instruments for cutting embryonic parts. Very find needles can be made by mounting 2-3 cm long tungsten wire (0.25 mm diameter) on a solid glass rod of 5 mm diameter and 15 cm length. The point is sharpened by dipping it in molten sodium nitrite ($NaNO_2$) in a porcelain or fine thread, finally with the help of a micro burner. These needles can be of varying thickness, shape and sizes, depending on requirements. One needs initial training to make good glass needles or to sharpen the tungsten needles.

Pipettes

In addition to ordinary Pasteur pipettes with rubber teats, one often needs pipettes of various shapes and sizes and micropipettes to handle small embryonic material. For this a 2 mm bore glass tubing is drawn out to 0.5 mm bore at the tip. A convenient teat is made by a 1-2 cm long 2 mm bore rubber tubing; on one side of it is inserted the glass tube while on the other is inserted a 5 mm length of solid glass.

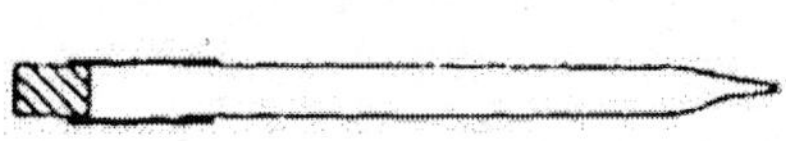

A Micropipette

Binocular

A zoom lens dissecting binocular microscope with a magnification of about 10x to 50x recommended for general use.

The experiments described in the following pages are relatively simple ones, which can be performed with ease. Our intention is to let the student handle the eggs and embryos, instead of studying embryology, only through prepared slides.

Other Equipment and Instruments Needed

1. Stereoscopic Binocular microscope, magnification range 15x-35x.
2. Microscope lamp
3. **Heat absorbing flask:** 250 cc round bottom flask filled with distilled water and supported by screw clamp on a vertical stand should be kept close to the microscope lamp, in order to absorb heat radiations from the latter.
4. Incubator
5. Finger howls
6. Petri dishes
7. Cavity slides
8. Coverslips (round)
9. Micro slides

10. Glass tubing, 3 mm diameter.
11. Glass rod (4 mm and 2 mm diameter).
12. Hypodermic syringe
13. Watch maker's forceps
14. Adjustable needle holders
15. Fine, sharp scalpels including a lancet type one
16. Bees wax
17. Low melting point paraffin wax
18. China marking pencils, red and, blue
19. Diamond point pencil
20. **Hair loops.** It used for handling embryos. These can be made by drawing a 5-6 mm soft glass tubing over a flame so that the smaller end has a diameter of about 1 mm. close the larger end of the tubing over a flame. The total length of such a piece should be about 10 cms. Take a hair and cut it into 2.5 cm long pieces. Using fine forceps, inset the two ends of the hair piece once at a time, into the open, narrow end of the tube and tip the same in soft, molten paraffin wax. Some wax will rise on the rube and get hardened on cooling and will hold the ends of the hair securely. To remove any wax, adherent to the hair loop, warm a slide on a flame, place it on a filter paper and gently touch the loop to the filter paper. Do not touch the loop to the warmed slide as the wax inside the tube will melt, releasing the ends of the hair.

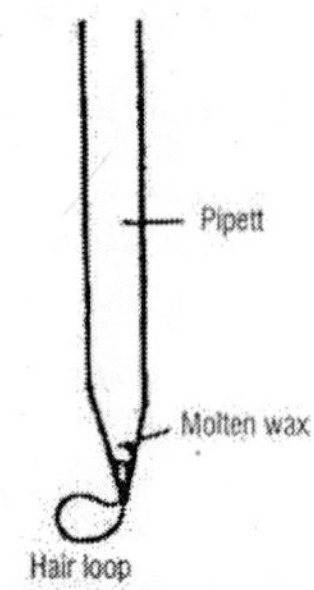

Hair loops

21. **Glass ball tips.** It can be made by drawing 6-7 mm soft glass rod over the flame. The drawn out narrow end should be broken and the pointed end should be held in the flame. A round glass ball will be formed at the tip. Constant spinning of the rod helps in having an almost perfectly round ball. Soine rods can be given a gentle curve while drawing them. These are very useful for moving embryos without injuring them.

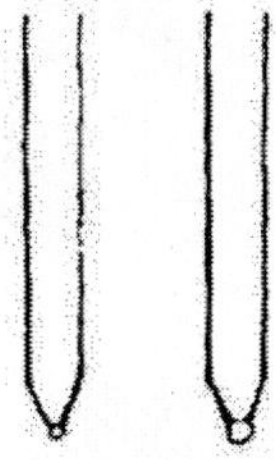

22. **Glass droppers.** It can be had readymade or can be drawn from soft glass tubing. Wide-mouthed pipettes (inside diameter 4 cm) are also needed.
23. **Candling apparatus.** For study of chick eggs it can be easily made. Fix a light socket and a switch on a board. To the light socket, fit a 100 watt bulb. Invert a large tin funnel over the bulb. Invert a large round tin over the bulb and the funnel. Cut a circular hole on the bottom of the tin and fasten a wire gauze below the hole. Cover the hole, from the outside, with a heavy black felt clothing in which a slit, slightly shorter than the length of an average egg, has been cut. The egg is placed over the slit in the, black cloth, the light is switched on and a strong beam of light will come up through tilt egg so that the blastoderm can be seen directly.
24. Sterilized absorbent cotton.

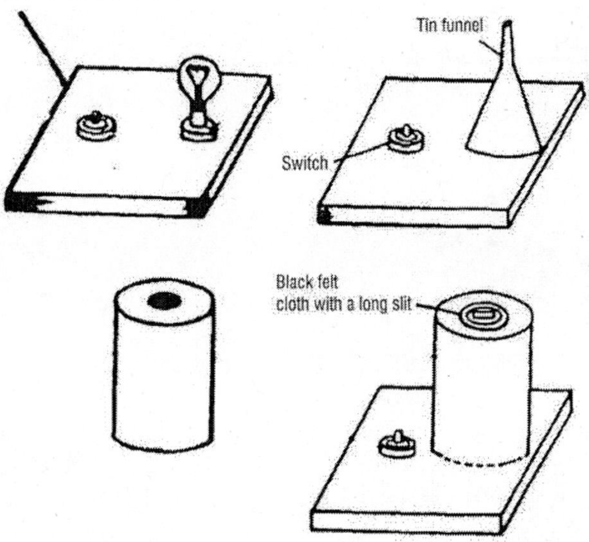

How to Clean Equipments

Before we discuss the experiments some fundamental pre-requisites must be understood. Before handling an egg or an embryo, the student must be aware that everything in the laboratory, including glassware, instruments and the hands of the worker i.e. absolutely clean and sterile, this is what is referred to as biologically clean. The idea is that the embryos must not be exposed to contamination by any chemical substances, living parasites or micro-organisms like bacteria or viruses. Following precautions are suggested for maintaining biological cleanliness.

1. Glassware must be thoroughly scrubbed with hot soap water and rinsed with running water for atleast 2 hours. It should then be rinsed with distilled water and air dried. Properly cleaned glassware should then be wrapped in, sterile paper and heat sterilized for 30 minutes. It should be kept wrapped in paper till required.
2. Instruments should always be washed, rinsed in distilled water and sterilized in an autoclave at. 15 pounds pressure for 30 minutes. Immediately before use, they should be immersed in 95% alcohol, those instruments which are to be used for operating living embryos should be of steel and not chromium plated ones. They should be kept in plastic envelopes, when not in use. Separate set of instruments should be used for handling dead embryos or for preserving eggs and embryos.
3. Dead or dying embryos must always be isolated and discarded at the earliest to avoid contaminating healthy ones. The latter should be briefly immersed in hypertonic salt solution which is reported to be an effective bacteriocidal agent.
4. Hands must be washed thoroughly at every step, as otherwise, formaldehyde or acid fumes, adhering, to the hands will contaminate the instruments and ultimately the embryos.

Glossary

Abdomen. That portion of the body cavity containing the greater part of the digestive system.

Acinus. Referring to a minute saclike structure.

Acrosome. The anterior extremity of the head of a spermatozoon. Derived from Golgi-complex, it contains proteolytic enzymes that dissolve egg envelopes and egg-plasma membrane during fertilization.

Adenohypophysis. The anterior lobe of the hypophysis (pituitary).

Adrenal. An endocrine gland near or upon the kidney.

Adrenergic. Relating to nerve fibres that liberate adrenalin.

Activation. Process of initiating development in the egg, normally achieved by the spermatozoon of the same species but also accomplished artificially (parthenogenesis); term also used to refer to stimulation of spermatozoon' to accelerated activity by chemical (fertilizin) means.

Adnexa. Extra embryonic structures (e.g., yolk sac) discarded before the adult condition is attained.

Aestivation. Reduced activity of some animals during the heat of the summer, opposed to term hibernation.

Agglutination. Cluster formation; a spontaneously reversible reaction of spermatozoa to certain chemical situations (e.g., egg water).

Albumen. A protein substance secreted by the walls of the oviducts around the egg of reptiles and birds.

Albumen sac. A two-layered ectodermal sac which encloses the albumen of the chick egg during early development of the embryo, separated for a time from the yolk by the vitelline membrane; later to release some of its contents into the amniotic cavity through the ruptured sero-amniotic connection.

Alecithal eggs. Literally meaning "without yolk", these eggs contain little or no yolk altogether.

Allantoin. A nitrogenous portion of the allantoic fluid.

Allantois. Sac-like organ of respiration and excretion found in reptiles, birds, and mammals and consisting of splanchnic mesoderm and evaginated hindgut endoderm.

Amnion. Thin, double membrane enclosing the embryos of some invertebrates and of reptiles, birds and mammals. In the vertebrate groups it is derived from the somatopleure.

Amniotic raphe. Region of junction of the amniotic folds as they encircle the embryo; Syn., sero amniotic or chorio-amniotic junction of raphe, amniotic umbilicus.

Amniogenic. Relating to the formation of the amnion.

Amniote egg. An egg which is isolated, from the environment by a near unpervious shell during embryonic development. It is self-contained and requires only oxygen from the outside.

Amphibia. A class of vertebrates, intermediate in many characters between fishes and reptiles, which live part of the time in water and part on land.

Amphiblastula. Double-structured blastula as in the Porifera (sponges).

Athphimixis. Mixing of germinal substances accomplished during fertilization.

Amplexus. The sexual embrace of the female amphibian by the male, a process which may (frogs and toads) or may not (urodeles) occur at the time of oviposition.

Anal plate. A thickening and invagination of mid-ventral ectoderm just poiterior to the primitive streat at the 14 hour stage which meets evaginating endoderm of the hindgut, later to be perforated as the proctodeum (anus). Syn., cloacal membrane.

Analogy. Similarity of parts in respect to function rather than to structure.

Anamniota. Forms which never develop an amnion, Cyclostomes, Fishes, Amphibia.

Anaphase. Phase of mitosis when the paired chromosomes are separating at the equatorial plate and begin to move toward the ends of the spindle.

Animal hemisphere. That half of the egg which contains the nucleus and active cytoplasm. It is most active in cell cleavage characterized ley less yolk. The other half of the egg is vegetal hemisphere.

Animal pole. The region of the egg where the polar bodies are formed region of telolecithal egg containing the nucleus and bulk of cytoplasm. Gives rise largely Co ectodermal derivatives. Syn., apical pole or hemisphere.

Anterior. Toward the head; head end. Syn, cephalic, cranial, rostral.

Aqueous humor. Fluid which fills the anterior and posterior chambers of the eye between the lens, probably derived from mesoderm.

Archenteron. Primitive gut found in the gastrula and communicating with the outside by the blastopore; precursor of the embryonic gut. Syn., Gastrocoel, Enteron.

Area opaca. Marginal opaque ring of the chick blastoderm which surrounds the area pellucida; less transparent because of the underlying adherent yolk.

Area pellucida. The central portion of the chick blastoderm, originally represented by the central cells, where the blastoderm is separated by a cavity from the underlying yolk and therefore appears to be more transparent.

Area vasculosa. The three-layered portion of the area opaca in which blood islands develop, appearing first posterior to the embryo and progressing forward, opposite the sides of the embryo, following the expansion of the mesoblast.

Area vitellina. Two-layered peripheral portion of the area opaca, itself subdivided into area vitellina interim and area vitellina externs, the latter including the zone of junction.

Asexual reproduction. Reproduction without union of gametes; generally no maturation divisions.

Aster. Aggregate of lines radiating from centrosome during animal mitosis.

Astray rays. Lines which make up the aster.

Autolysis. A process of disintegration of a cell by its own lyosomal enzymes.

Axis of the embryo. Line (imaginary) representing the antero-posterior axis of the future embryo.

Balfour's law. The velocity of segmentation in any part of the ovum is, roughly speaking, proportional to the concentration of the protoplasm there; and the size of the segments is inversely proportional to the concentration of the protoplasm.

Bilateral symmetry. An anatomical arrangement in which an organism has two equal but opposite right and left halves when it is cut lengthwise in the middle vertical plane.

Biogenetic law. The embryos of higher species tend to resemble embryos of lower species in certain respects but they are never like the adults of lower species. Embryonic development is a gradual deviation from the more general (phylogenetic) to the more specific characters of the individual species. Not to be confused with the recapitulation theory.

Blastema. An indifferent group of cells about to be organized into definite tissue; the newly formed cells covering a cut surface, functional in regeneration of tissues.

Blastocoel. The cavity of the blastula. Syn., segmentation or subgerminal cavity.

Blastocyst. Mammalian blastula, containing large blastocoel. Syn., blastodermic vesicle.

Blastoderm. The living portion of the egg from which both the embryo and all of its membranes are derived. The cellular blastodisc.

Blastodisc. The disc of protoplasm within which cleavage occurs in eggs of fishes, reptiles, or birds. Found in telolecithal eggs having meroblastic or discoidal cleavage. Syn., germinal disc.

Blastomere. A cellular unit of the developing egg or early embryo, prior to the time of gastrulation. Smaller blastomeres are micromeres; intermediate are mesomeres; larger are macromeres, where there is great disparity in size.

Blastopore. The opening of the archenteron (gastrocoel) to the exterior, occluded by the yolk plug in amphibian embryos; consisting of a slit-like space between the elevated margin of the blastoderm and the underlying yolk of the chick egg; represented in the amniota as the primitive streak.

Blastopore. Dorsal lip of the region of first involution of cells in the amphibian gastrula; general area of the organizer; original grey crescent area; the cells which turn in beneath the potential central nervous system (Amphioxus) and form the roof of the archenteron. Syn., germ ring.

Blastopore. Ventral lip of region of blastopore opposite the dorsal lip; region which gives rise to ectoderm overlying the central nervous system of Amphioxus or to the peristomial mesodem of the frog, Syn., germ ring.

Blastula. The stage in embryonic development between the appearance of distinct blastomeres and the end of cleavage (i.e., the beginning of gastrulation); a stage generally possessing a primary embryonic cavity or blastocoel; invariably monodermic.

Blood islands. Pre-vascular groups of mesodermal cells found in the splanchnopleure, from which will arise the blood vessels and corpuscles.

Branchial cleft. An opening between branchial arches formed by invaginating head ectoderm and evaginating pharyngeal endoderm (pouch) through which water passes from the pharynx to the outside of the frog. Syn., gill cleft or slit, some visceral clefts.

Branchial groove. Ectodermal invagination anterior or posterior to visceral arch which joins branchial pouch to form branchial cleft, in most instances.

Bud. An undeveloped branch, generally an anlage of the appendage. (e.g., limb or wing bud).

Bulbus arteriosus. The most anterior division of the early, tubular, embryonic heart which leads from the ventricle to the truncus arteriosus.

Bursa fabricii. Endodernial cavity derived from the posterior portion of the embryonic cloaca and communicating with the dorsal part of the proctoderm at the level of the urodeal membrane.

Cell. The structural and functional unit of plant and animal organization or protoplasm surrounded by a lipoproteinaceous plasma membrane and containing one or more nuclei.

Cell lineage. The study of the origin and fate of specific blastomeres in embryonic development. Syn., cytogeny.

Cell theory. The body of any living organism is composed of structural and functional units, the primary agents of organization called cells. Each cell consists of a nucleus and its sphere of influence, consisting of cytoplasm, generally circumscribed by a membrane. "Omnis cellula e cellula".

Central cells. Cells of the chick blastoderm which are bounded on all sides in surface view. May be connected with each other below by contiguous syncytial protoplasm. Surrounded by marginal periblast.

Central periblast. The protoplasmic layer beneath the segmentation cavity of the early fish blastula.

Centriole. The granular core of the centrosome.

Centrolecithal. Pertaining to eggs with yolk accumulated in centre of cell, e.g., in arthropods.

Centromere. Region of constriction of chromosomes. It holds sister chromatids together and spindle fibres attach to it, cell division.

Centrosome. The granule (centriole) and the surrounding sphere of rays (centrosphere) which function as kinetic centers in mitosis. The center of the, aster which does not disappear when the astral rays disappear.

Cerebral flexure. Flexures of the head region, including the cranial, pontine, and the cervical flexures.

Cerebral peduncles. Longitudinal tracts in the floor of the mesencephalon.

Cervical flexure. Extensive transverse flexure of the chick embryo occurring at the level of the junction of the head and trunk about the end of the second day.

Cervical sinus. Depression between the third and fourth oral clefts (pig) and the body.

Cetacea. An order of completely aquatic, mostly marine, mammals of the subclass Eutheria, including the whales, dolphins, and porpoises.

Chalaza. Twisted opalescent chord of heavy albumen immediately surrounding the vitelline membrane of the chick egg, which helps to hold the yolk in position.

Chiasma. Describing the crossing of fibers.

Chimera. Compound embryo generally derived by grafting major portion of two embryos, usually of different species; may be derived by abnormal chromosomal distributions in cleavage after normal fertilization.

Chorda-mesoderm. Region of the late (amphibian) blastula arising from the grey crescent area, which will give rise to the notochord and mesoderm and will, if transplanted, induce the formation of secondary medullary folds.

Chorio-allantois. A common membrane formed by the fusion of the inner wall of the chorion and the outer wall of the allantoid (both mesoderm) in the chick embryo.

Chorion. An extra-embryonic membrane of the chick embryo which develops from the stomatopleure as a corollary of the amnion, and which encloses both the amnion and the allantois; consists of inner mesoderm and outer ectoderm. Syn., serosa, false amnion. The term also used to designate the tough membrane deposited by cells of the ovarian follicle on fish eggs.

Choroid fissure. An inverted groove in the optic stalk whose lips later close around blood vessels and nerves that enter the eyeball.

Chromosome. The chromatic or deeply staining bodies derived from the nuclear network and containing a matrix and one or more chromonemata during the process of mitosis; bodies found in all somatic cells of the normal organism in a number characteristic of the species; bearers of the gene.

Cicatrix. The thin, non-vascular side of the ovarian follicle where rupture will occur to allow the extrusion of the egg into the body cavity in vertebrates. Syn., stigma.

Cleavage. The mitotic division of tie egg resulting in blastomeres. Syn., segmentation.

Cleavage, accessory. Cleavage in peripheral or deeper portions of the (chick) germinal disc caused by supernumerary sperm-nuclei following (normal) polyspermy.

Cleavage, asymmetrical. Extremely unequal divisions of the egg as in the Ctenophore.

Cleavage, bilateral. Cleavage in which the egg substances are distributed symmetrically with respect to the median plant of the future embryo.

Cleavage, determinate. Cleavage in which certain parts of the future embryo may be circumscribed in certain specific (early) biastomeres; cleavage which produces blastomeres that are not qualitatively equipotential, i.e., when such blastometes are isolated they will not give rise to entire embryos. Syn., mosaic development.

Cleavage, dexiotropic. Cleavage resulting in a right-handed production of daughter blastomere(s), as in spiral cleavage.

Cleavage, equatorial. Cleavage at right angles to the egg axis, opposed to vertical or meridional; often the typical third cleavage plane. Syn., latitudinal or horizontal.

Cleavage, holoblastic. Complete division of the egg into blastomeres, generally equal in size although not necessarily so (e.g., Amphioxus). Syn., total cleavage.

Cleavage, indeterminate. Cleavage resulting in qualitatively equipotential blastorneres in the early stages of development; when such blastomeres are isolated from each other they give rise to complete embryos. Opposed to mosaic development. Syn., regulatory development.

Cleavage levotropic. Cleavage resulting in left-handed or counter-clockwise production of daughter blastomere(s) as in some cases of spiral cleavage.

Cleavage, meridional. Cleavage along the egg axis, opposed to equatorial; generally the first two cleavages of any egg. Syn., vertical.

.Cleavage, meroblastic. Cleavage restricted to peripherally located protoplasm, as in the chick egg. Syn., discoidal cleavage.

Cleavage nucleus. The nucleus which controls cleavage. This may be the syngamic nucleus of normal fertilization; the egg nucleus of parthenogenetic or gynogenetic eggs; or the sperm nucleus of androgenetic eggs.

Cleavage path. Path taken by the syngamic nuclei to the position awaiting the first division.

Cleavage, radial. Holoblastic cleavage which results in tiers of cells.

Cleavage, spiral. Cleavage at an oblique angle with respect to the egg axis so that the resulting blastomeres (generally micromeres) lie in an interlocking fashion within the furrows of the original blastomeres, due to intrinsic genetic factors. (e.g., Mollusca).

Cleavage, superficial. Cleavage around the periphery of centolecithal eggs. Syn., peripheral cleavage.

Cleidoic. Referring to eggs, such as those of birds that are insulated from the environment by albumen, membranes, and shell.

Coelom. Mesodermal cavity from the walls of which develop the gonads; cavity subdivided in higher, forms into pericardial, pleural, and peritoneal cavities.

Coitus. Copulation of male and female, term generally used in connection with mammals.

Competence. Ability of an embryonic area to react to a stimulus (e.g., evocator).

Concrescence. The coming together of previously separate parts (cells) of the embryo generally resulting in a piling up of parts, one of the corollaries of gastrulation in many forms.

Cone, fertilization. Conical projection of cytoplasm from surface of egg to meet spermatozoon which is to invade egg cortex. Cone makes contact and then draws sperm into egg. Not universally demonstrated or seen in frog, but seen in starfish (chambers). Syn., exudition cone.

Cones of growth. The enlarged outgrowth of the neurobla

Convergence. Movement of groups of embryonic cell.

Copulation path. The second portion of sperm migration path through the egg toward the egg nucleus, when there is any deviation from the entrance or penetration path; the path of the spermatozoon which results in syngamy.

Coracoid. Shaped like a crow's beak.

Corona radiata. Layer of follicle cells which immediately surround the egg of the bird or mammal, and which cells are elongated and acquire intercellular canals radiating outwardly from the egg to the surrounding theca. Syn., follicular epithelium.

Corpus albicans. Scar tissue over the point of rupture of an ovarian follicle (mammal) following absorption of the corpus luteum.

Corpus haemorrhagicum. Blood clot found in recently ruptured ovarian follicle of the mammal.

Corpus luteum. A yellow granular substance derived from the zona granulosa found in the empty follicular vesicle after ovulation in mammals; endocrine function relative to gestation and ovulation. Substance called lutein.

Cortex. The Outer portion of an organ part or egg.

Cotyledonous placenta. A chorion, as in the cow and other ruminants, which is made up of brush-like groups of villi. (Opposed to diffuse and intermediate placentae).

Crescent, gray. Crescentic area between the original animal and vegetal pole regions on the surface of the frog's egg, gray in colour because of the migration of pigment away from the area and toward the sperm entrance point; region of presumptive chorda-mesoderm, future blastopore and anus.

Crescent, yellow. Crescentic area on the surface of the Ascidian egg, yellow in colour, which will give rise to the mesoderm of the embryo.

Crescent, mesodermal. Region of the Amphioxus egg seen in the 4-cell stage, due to pigment migration, which will give rise to the embryonic mesoderm.

Cross-fertilization. Union of gametes produced by different individuals which, if they are of different species, may produce hybrids.

Crossing over. Mutual exchange of portions of allelomorphic pairs of chromosomes during the process of synapsis in maturation.

Crypts. Depressions found in the uterine wall for the reception of chorionic cotyledons.

Ctenoid. A type of scale featuring tooth like projections.

Cumulus oophorus. The aggregation of cells which immediately surrounded the mammalian egg within its Graafian follicle.

Cycle, sexual. Cyclic breeding activity, most evident in the female mammal associated with definite changes in the genital and endocrine systems.

Cytokinesis. Division of the cytoplasm of a cell during cell division.

Cytology. The study of cells.

Cytolysis. The breakdown of the cell indicated by a dispersal of formed components.

Cytoplasm. The material of the cell exclusive of the nucleus; protoplasm apart from nucleoplasm.

Cytotrophoblast. The inner or cellular layer of the trophoblast.

Decidua. The portion of the uterine wall cast off at the time of parturition.

Decidua basalis. That portion of the uterine wall to which the placenta is attached. Syn., decidua scrotina.

Decidua capsularis. The portion of the uterine mucosa and epithelium which covers the mammalian blastocyst opposite of the uterine mucosa and epithelium which covers the mammalian blastocyst Opposite the placenta. Syn., decidua reflexa.

Decidua vera. The portion of the uterine wall aside from that associated with either the decidua basalis or the decidua capsularis, but which will eventually be in contact with the decidua capsularis with the expansion of the embryo.

Deciduate. The type of placentation which is characterized by the destruction of material (uterine) tissue, and hemorrhage at parturition. True placenta.

Dedifferentiation. Disappearance or loss of specialized or differentiated cell structures reverting them to a differentiated state like the embryonic cells.

Delamination. A separation (of cell layers) by splitting, a process in mesoderm formation.

Dermatome. The outer unthickened wall of the somite which gives rise to the dermis. Syn., cutis plate.

Determination. Process of development indicated when a tissue, whether treated as an isolate or a transplant, still develops in the originally predicted manner.

Determination of sex. Method by which the sex of the unborn is revealed. Not to be confused with sex determination which is generally achieved at the time of fertilization.

Deutoplasm. Yolk or secondary food substances of the egg, non-living.

Development. It is a process by which organisms undergo progressive, orderly and gradual changes hi structure as well as in physiology during their entire life history.

Differentiation. Increase in the visible an invisible complexities of the organism progressive diversification leading to histogenesis; the process which changes a simple to a complex organism.

Diffuse placenta. Placenta in which the individual villi are scattered over the intercotyledonary areas (e.g., Giraffe).

Discoblastula. A disc-shaped blastula found in cases of discoidal (meroblastic) cleavage. (e.g., Cephalopoda and the Chick).

Discoidal placenta. Placenta developing on only one side of the blastocyst (e.g., Rodents) in the general shape of a button or disc. Opposed to zonary placenta.

Discus proligerus. The oocyte of the mammal is attached to the inner wall of the follicle by its neck which, along with the cells surrounding the ovum, are together known as the discus proligerus.

D-quadrant. One of the four early blastomeres of the annelid embryo which has the prospective function of giving rise chiefly to mesoderm.

Dyads. Aggregations of chromosomes consisting of two rather than for (tetrad) parts, term used to describe condition during maturation process.

Ectoderm. The outermost layer of the didermic gastrula. Syn., epiblast.

Egg, alecithal. Eggs with little or no yolk. Literally means without yolk.

Egg, cleidoic. Eggs such as those of reptiles, birds, and oviparous mammals, which are covered by a protective shell.

Egg, ectolecithal. An egg having the yolk around the formative protoplasm, opposed to centrolecithal.

Egg envelope. Material enveloping the egg but not necessarily a part of the egg, such as vitelline membrane, chorion, jelly, albumen.

Egg, giant. Abnormal polypoid condition where chromosome completes are multiplied, resulting in giant cells and embryos.

Egg, homolecithil. Egg (e.g., Mammal) with little yolk which is scattered throughout the cytoplasm.

Egg, isolecithal. Eggs with homogeneous distribution of yolk, may be isolecithal, alecithal, or homolecithal.

Egg jelly. The mucin covering deposited on the amphibian egg as it passes through the oviduct.

Egg, macrolecithal. Egg with large amount of yolk, generally telolecithal.

Egg membranes. Includes all egg coverings such as vitelline membrane, chorion, and the tertiary coverings.

Egg, microlecithal. Egg with small amount of yolk. Syn., meiolecithal, oligolecithal.

Egg, telolecitbal. Egg with large amount of yolk concentrated at one pole.

Egg tooth. A mammiform hard structure with pointed nipple, found on the dorsal side of the tip of the upper jaw at the time of hatching of the chick embryo, with the sole function of aiding the chick in the hatching process.

Egg water. Watery extract of materials diffusing from living eggs, presumably the Fertilizin of Lillie. Syn., egg water extract.

Emboitement. The preformationist theory of Bonnet and others based on the idea that the ovary of the first female contained the miniatures of all subsequently existing human beings. Syn:, encasement theory.

Embryo. Any stage in the ontogeny of the fertilized egg, generally limited to the period prior to independent food-getting. The stage between the second week and the second month of the human embryo.

Embryonic disc. The portion of the early mammalian embryo where the ectoderm and endoderm are in close contact with each other.

Embryonic field. Region of formative processes within the embryo, larger than the area of ultimate realization of structures concerned.

Embryonic knob. The inner, trophoblastic, cell mass of the mammalian embryo (ectodermal).

Embryonic membranes. Refers to the amnion, chorion (serosa), allantois and the yolk-sac splanchnopleure of the chick embryo.

Embryonic shield. The region of the fish blastoderm which will give rise to the embryo proper; the area of ectoderm and adherent endoderm out of which the mammalian embryo will develop.

Embryotrophy. The means or the actual nourishment of the mammalian embryo.

Enterocoel. The cavity or pouch within the mesoderm just formed by evagination of the gut (enteron) endoderm as in Amphioxus. Syn., gut pouch, coelomic pouch, archenteric pouch.

Enterori. The definitive gut of the embryo, always lined with endoderm.

Entobronchi. Six, dorsal outgrowths of the mesobrorichus on the sixth day (chick) related to the respiratory system of the adult.

Ento-mesoderm. Refers to the portion of the invaginating blastoporal lips which will induce the formation of medullary fields in the amphibian embryo.

Entrance cone. The temporary depression on the surface of the egg following the entrance of the spermatozoon.

Entypy. A method of amnion formation in the mammal in which the trophoblast above the embryonic knob is never interrupted; a method of gastrulation (rodents) wherein the endoderm comes to lie externally to the amniotic ectoderm.

Epiblast. Outermost layer of the early embryo from which the various germ layers may be derived.

Epiboly. Growing, spreading or flowing over; process by which the rapidly dividing animal pole cells or micromeres grow over and enclose the vegetal pole Material.

Epigenesis. Development of systems starting with primitive, homogeneous, lowly organized condition and achieving great diversification.

Epimere. The most dorsal mesoderm, that lying on either side of the nerve and notochord, which gives rise to the somites. Syn., axial mesoderm.

Equatorial plate. The latest view of the chromosomes, lined up on the mitotic spindle, prior to any anaphase movement.

Evagination. The growth from any surface outward. Exocoel the cavity within mesodem beyond the limits of the (chick) embryo, continuous with the coelom. Syn., extra embryonic body cavity; coelom; seroamniotic cavity.

Exogastrula. Gastrulation modified experimentally by abnormal conditions so that invagination is partially or totally hindered and there remains some mesendoderm not enclosed by ectoderm.

External gills. Outgrowths of (amphibian) branchial arches which function as temporary (anura) or permanent neotonic (Urodela) respiratory organs.

Extra embryonic. Refers to structures apart from the embryonic body, such as the membranes.

Fertilization. Activation of the egg by sperm and syngamy of the pronuclei; union of male and female gamete nuclei.

Fertilization cone. A conical projection of cytoplasm from the surface of the egg to meet the spermatozoon which is to invade the egg; makes contact and then draws the spent' into the egg.

Fertilization membrane. A non-living membrane seen to be distinct from the egg shortly after fertilization very probably the vitelline membrane elevated off of the egg.

Fertilizin. Chemical substance in the cortex of the mature egg apparently necessary for normal fertilization. Syn., sperm iso-agglutinin.

Fetus. Embryo after it has attained definite characteristics of the species; the embryo of the mammal when it has attained sufficient development to survive independently of the mother.

Follicle, graafian. The follicle of the mammalian ovary, including a double-layered capsule; the membrana granulosa, discus proligerus, corona radiata, and follicular fluid. -

Gamete. A differentiated (maturated) germ cell, capable of functioning in fertilization e.g., sperm, or eg cell; germ cell.

Gametogenesis. The process of developing and maturing germ cells.

Gastrula. The didermic embryo, possessing a newly formed cavity the gastrocoel or archenteron.

Genital ridge. Initial elevation for the development of the external genitalia; paired, meosdermal thickenings between the mesonephros and the dorsal mesentery of all vertebrates, which are the gonad primordia.

Germ. The egg throughout its development, or at any, stage.

Germ cell. A cell capable of sharing in the reproductive process, in contrast with a somatic cell. E.g. sperm or egg cell. Syn.; gamete.

Germ layer. A more-or-less artificial spatial and histogenic distinction of cell groups beginning in the gastrula stage consisting of the ectoderm, endoderm and mesodermal layers. No permanent and clear-cut distinction, as shown by transplantation experiments.

Germ plasm. The hereditary material, generally referring specifically to the genotype. Opposed to somatoplasm.

Germ ring. Ring of cells showing accelerated Mitotic activity, generally a synonym for the lips of the blastopore. The rapidly advancing cells in epiboly.

Germ wall. The syncytial periblast of the chick embryo which is being added to the central,, well-defined, cellular blastoderm; the advancing peripheral boundary of the chick blastoderm.

Germinal disc. The small protoplasmic disc on the surface of the huge yolk mass of the (hen's) egg.

Germinal epithelium. The peritoneal epithelium out of which the reproductive cells of both the male and female develop. Syn., germinal ridges, gonadal ridges.

Germinal vesicle. The pre-maturation nucleus of the egg.

Gestation. Period of carrying the young (mammal) within the uterus.

Growth. A developmental increase in total mass of protoplasm at the expense of raw materials; an embryonic process generally following differentiation.

Gynandromorph. Condition where part of an animal may be male and another part female, not to be confused with hermaphroditism.

Gynogenesis. Development of sperm activated egg, but without benefit of the sperm nucleus.

Haploid. Having a single set of chromosomes not appearing in allelomorphic pairs, as in the mature gametes. Opposed to diploid, or the condition in somatic cells.

Hatching. The beginning of the larval life of the amphibian, accomplished by temporarily secreted hatching enzymes which aid the embryo to escape its gelatinous capsule; the process of emergence of the chick embryo from its shell; involving critical changes in the body such as closing of the umbilicus, contraction of the ductus arteriosus, and drying up of the membranes. Hatching of chick aided by beak tooth.

Head fold. An anteriorly projecting semi-circular fold of the medullary plate of the chick appearing first at about 21 hours. The fold involves both ectoderm and endoderm, and is found also in mammals. Becomes head process.

Head fold amnion. Fold of the amnion which envelops the head of the chick embryo.

Head process. Originally a faint line of cells anterior to Hensen's node and continuous with the primitive streak but slightly out of line with the streak, the first indications of the embryonic sac, later to enlarge as an actual anterior projection of ectoderm and endoderm.

Hemiplacenta. The chorion, the yolk sac, and generally the allantois which together serve as an organ of nutritional supply to the uterine young of marsupials.

Hensen's knot. The anterior termination of the primitive streak, consisting of a knot of cells possibly homologous to the dorsal lip of the blastopore of the amphibia. Syn., Hensen's node, primitive knot or node.

Hermaphrodite. An individual capable of producing both spermatozoa and ova.

Hermaphrodite, protandrous. Male elements mature prior to female elements in hermaphrodite.

Hermaphrodite, protogynous. Female elements mature prior to male elements in hermaphrodite.

Hertwig's law. The nucleus tends to place itself in the center of its sphere of activity; the longitudinal axis of the mitotic spindle tends to lie in the longitudinal axis of the yolk-free cytoplasm of the cell.

Heteroplastic transplant. A graft from one individual to another, generally of a different species.

Heterotaxis. Exceptional case wherein the chick embryo rotates so as to lie on its right side, often associated with sites inversus viscerum; may be induced experimentally.

Heterozygous. Condition where the zygote is composed of gametes bearing allelomorphic genes. Opposed to homozygous.

Hibernation. Spending the cold (winter) period in a state of reduced activity.

Histogenesis. The development of the tissues.

Hoffman's nucleus. Group of neuroblasts at the external margin of the white matter just above the ventral roots in the spinal cord of the chick; extends the length of the cord.

Homoiothermal. An adjective which refers to condition where the temperature of the body of the organism is under the control of an internal mechanism body temperature regulated. Opposed to poikilotherinal.

Homoplastic. Refers to a graft to an organism of the same species, or even to another position on the same individual. Syn., autoplastic.

Homunculus. Imaginary human form presumed to exist within the egg or sperm; preformationism.

Hybrid. A successful cross between different species. E.g., horse and ass give a (sterile) mule.

Hyperplasia. Overgrowth; abnormal or unusual increase in elements composing a part.

Hypochordal rod. A transitory string of cells constricted off between the dorsal wall of the midgut and the notochord of the amphibian embryo, between the level of the pancreas and the tail and disappearing before hatching time. Syn., subnotochordal rod.

Hypomere. The most ventral segment of mesoderm out of which develop the somatopleure, splanchnopleure, and coelom. Syn., lateral plate mesoderm.

Hypoplasia. Undergrowth or deficiency in the elements composing a part.

Idiozome. The material out of which the acrosome is formed during the metamorphosis of spermatid to spermatozoon.

Implantation. The process of adding, superimposing, or placing a graft within a host without removal of any part of the host; the attachment of the mammalian blastocyst.

Incubation. The application of regulated heat to a developing egg.

Indeciduate placenta. The type of placenta in which each villus simply fits into a crypt, as a plug fits into a socket, from which it is withdrawn at birth without serious hemorrhage or destruction of maternal tissues. Syn., nondeciduate.

Induction. Successive and purposeful influences which bring about morphogenetic changes within the embryo.

Infundibulum of the brain. Funnel-like evagination of the floor of the diencephalon which, along with the hypophysis, will give rise to the pituitary gland of the adult.

Inner cell mass. The spheribal cells on the inner side of the mammalian blastocyst.

Insemination. The process of impregnation; to fertilize.

Interauricular septum. A longitudinal sheet of (mesodermal) tissue which grows ventrally from the roof of the auricular chamber to divide it into right and left halves.

Interkinesis. Resting stage between mitotic divisions.

Intermediate cell mass. The narrow strip of mesoderm which-, for a time, joins the- dorsal epimer with the ventral hypomere, being made up of a dorsal portion continuous with the dorsal wall of the somite and the somatic mesoderm and a ventral portion continuous with the ventral wall of the somite and the splanchnic mesoderm. Source of origin of the excretory system. syn., nephrotorne or middle plate.

Invagination. The folding or inpushing of a layer of cells into a preformed cavity, as in one of the processes of gastrulation. Do not confuse with involution.

Involution. The rolling inward or the turning in of cells over a rim, as in the gastrulation of the chick embryo.

Isogamy. Similar gametes, without differentiations into spermatozoa and ova.

Karyoplasm. Protoplasm within the confines of the nucleus.

Kupffer's vesicle. A cavity at the posterior end of the early fish embryo presumed to represent a rudiment of the neurenteric canal.

Labio-scrotal swellings. The primordia of the labia majora (female) or of part of the scrotum (male) in mammals.

Lanugo. The fine hairy covering of the fetal mammal.

Larva. Stage in development when the organism has emerged from its membranes and is able to lead an independent existence, but may not have completed its development. Generally (except in cases of neotony or paedogenesis) larvae cannot reproduce.

Latebra. The bulbous portion of the flask-shaped mass of white yolk located toward the center of the chick egg, connected by a neck of similar material with the nucleus of Pander.

Lateral amniotic folds. Folds of the amnion extending up over the sides of the chick embryo, developing as corollaries to, or in consequence of, the head and tail amniotic folds.

Lateral line organs (or system). A line of sensory structures along, the side of the body of fishes and amphibia, generally embedded in the skin and innervated by a branch from the vagus ganglion. presumably concerned with the recognition of low vibrations in water. Appears first at about 4 mm stage in the frog embryo. Syn., ramus lateralis.

Lateral plates or lateal plate mesoderm. The lateral mesohlast within which the body cavity (coelom and exocoel) arises. Syn., lateral mesoderm.

Lateral rotaticn. Twisting which begins at the anterior end of the chick embryonic body and extends posteriorly until the embryo is lying on its left side.

Lens placode. The early thickened ectodermal primordium of the lens.

Lens. A thickening in the head ectoderm opposite the optic cup at about the time of hatching in the frog embryo and at the 26 somite stage in the chick, which becomes a opotic cup as a lens. Inner surface convex substance fibrous.

Lenticular zone. The portion around the rim of the optic cup adjacent to the pupil, separated from the retinal zone in later (chick eye) development by the ora serrata.

Liquor folliculi. The fluid of the mammalian Graafian follicle into which the matured ovum is freed wen the discus proligerus is broken, finally to be liberated from the ovary at the time of ovulation. Contains hormones.

Lutein. The yellow coloured material contained in the cells which fill the empty mammalian Graafian follicles.

Macromere. Larger of blastomeres where there is a conspicuous size difference.

Maternal placenta. Uterine mucosal portions of/the typical mammalian placenta.

Maturation. The process of transformation of a primordial germ cell (spermatogonium or oogonium) into a functionally mature germ cell, the process involving' two special divisions, one of which is always meiotic. -Divisions known as equational and reductional.

Megaspheres. Yolk masses which are cut off from the floor of the subgertninal cavity of the chick blastoderm and which lie within this cavity.

Meiosis. A process of nuclear division found in the maturation of germ cells, involving a separation of members of paits of chromosomes. Syn., reductional division.

Membrana granulosa. The layers of follicle cells which hound the mammalian follicular cavity.

Menstruation. Process in Primates caused by decrease in secretion of estrogen and progestrone, involving loss of endometrium and some hemorrhage.

Mesenchyme. The form of embryonic mesoderm or mesoblast in which migrating cells unite secondarily to form a syncytium or network having nuclei in thickened nodes between intercellular spaces filled with fluid; often derived from mesothelium.

Mesendoderm. Newly formed layer of the (urodele) gastrula before there has been any separation of endoderm and mesoderm. Syn., mesentoderm.

Mesentery. Sheet of (mesoderm) tissue generally supporting organ systems. E.g., mesorchium, mesocardium.

Mesobronchus. Division of the primary (chick) lung, parts of which give rise to air sacs.

Mesocardium. The mesentery of the heart; may be dorsal, ventral, or lateral.

Mesocolon. That portion of the embryonic dorsal mesentery which supports the colon.

Mesoderm. The third primary germ layer developed in point of time, may be derived from endoderm in some forms and from ectoderm in others.

Mesogastrium. Dorsal and ventral mesenteries which support the developing gut and its derivatives.

Mesomere. Cell of intermediate size where there are conspicuous size differences in an early embryo; also refers to Intermediate cell mass, Intermediate mesoderm.

Mesometrium. Attachment of the uterus to the coelomic wall.

Mesorchium. Mesentery (mesodermal) which surrounds and supports the testis to the body wall.

Mesotheoium. Epithelial layers or membranes of mesodermal origin.

Mesovarium. Mesentery (mesodermal) which suspends the ovary from the dorsal body wall.

Metamerism. Serial segmentation, as seen in the nervous, muscular and circulatory systems.

Metamorphosis. The end of the larval period of amphibia when growth is temporarily suspended. There is autolysis and resorption of old tissues and organs such as gills, and the development of new structures such as eyelids and limbs; changes in structure correlated with changes in habitat from one that is aquatic to one that is terrestrial; change in structure without retention of original form, as in change from spermatid to speramatozoon.

Micromere. Smaller of cells when there is a conspicuous difference in size, characteistic of Annelids and Molluscs.

Micropyle. An aperture in the egg covering through which spermatozoa may enter; in such eggs the only possible point of insemination. E.g., many fish eggs.

Milk-ridge. A band of tissue between the somites (dorsally) and the level of the heart, liver, and mesonephros (ventrally) in the pig embryo, which gives rise to the mammary glands; first seen at 12 mm and (well developed by) 15 mm.

Mitosis. Cytoplasmic division Involving a nucleus and spindle apparatus and equal distribution of chromosomes.

Mitotic index. The proportion of the dividing cells in any tissue and at any specified time.

Monospermy. Fertilization accomplished by only one sperm. Opposed to polyspermy.

Morula. A spherical mass of cells, as yet without segmentation cavity.

Myelocoel. Cavity of the myelencephalon; fourth ventricle.

Myoblasts. Formative cells within the myotome or muscle plate which will give rise to the true striated muscles of the adult.

Myocardium. The muscular part of the heart arising from the splanchnic mesoblast.

Myotome. The thickened primordium of the muscle found in each somite. Syn., muscle plate.

Nebenkern. Cytological structure near the nucleus of the early spermatid.

Neotony. Condition of many urodeles and of experimentally produced (thyroidless) anuran embryos where the larval period is extended or retained, i.e., the larvae fail to go through normal metamorphosis. e.g., Axolotl, Necturus.

Neural crest. A continuous cord of ectodermally derived cells lying on each side in the angle between the neural tube and the body ectoderm, separated from the ectoderm at the time of closure of the neural tube and extending from the extreme anterior to the posterior end of the embryo; material out of which the spinal and possibly some of the cranial ganglia develop, and related to the development of the sympathetic ganglia by cell migration.

Neural fold. Elevation of ectoderm on either side of the thickened and depressing medullary plate; folds which close dorsally to form the neural tube. Syn., medullary folds.

Neural groove. The sinking in of the center of the medullary plate to form a longitudinal groove, later to be incorporated within the neural tube (spinal cord). Syn., medullary groove.

Neural plate. Thickened broad strip of ectoderm along the future dorsal side of all vertebrate embryos, later to give rise to the central nervous system. Syn., medullary plate.

Neural tube. The tube formed by the dorsal fusion of the neural folds, the rudiment of the nerve or spinal cord.

Neurenteric canal. The posterior neuroccel where it is connected with the closing blastopore and posterior enteron of the amphibian; the large common nerves and enteric chamber of Amphioxus; the Kupffer's vesicle of the fish embryo; possibly the primitive pit of the chick embryo.

Neuroblasts. Primitive or formative nerve cells, probably derived (along with epithelial and glia cells) from the germinal cells of the three day chick embryo neural tube.

Neurocoel. The cavity of the neural tube, formed simultaneously with the closure of the neural folds, Syn., central canal, neural canal.

Neurocranium. The dorsal portion of the skull associated with the brain and sense organs.

Neuropore. A temporary opening into the neural canal due to a lag in the fusion of the neural folds at the anterior extremity; permanent in Amphioxus and in the vicinity of the epiphysis of higher vertebrates. There may be a temporary posterior neuropore in some, even in the chick.

Notochord. Rod of vacuolated cells representing the axis of all vertebrates, found beneath the neural tube and dorsal to the archenteron. Origin variable or doubtful, in most cases thought to be derived from or simultaneously with the endoderm.

Notochordal canal. An exaggerated primitive pit in some mammals which extends into the head process. May be homologous to neurenteric canal.

Notochordal sheath. Double mesodermal sheath around the notochord consisting of an outer elastic sheath developed from superficial chorda cells and an inner secondary or fibrous sheath from chorda epithelium.

Nucleolus. The body generally within the nucleus which has no affinity for chromatin dyes. But stains with acid or cytoplasmic dyes. Function unknown Syn., plasmosorn.

Nucleus of pander. The plate of white yolk beneath the blastodisc of the chick egg.

Nucleus, red. Synaptic center of the mid-brain which acts as a coordinating pathway for synergic type of muscular control.

Omphalomesenteric artery. Originally paired vessels originating from the dorsal aorta of Lhe ekrck embryo later to fuse as result of formation of the dorsal mesentery; function to carry blooMiducto the extra embryonic vitelline vessels of, "(lie chick embryo. Syn., vitelline artery.

Omphalomesenteric veins. The viteiline veins as they enter the body of the chick embryo at the level of the anterior intestinal portal; united posterior to the sinus venosus as the meatus venosus. Venosus rings join these vessels dorsal to the gut on the third, and then the fourth days to become a single curved vessel emptying through the liver into the meatus venosus. The vitelline vein is continuous with the omphalomesenteries, but is found on the yolk, and is therefore extra-embryonic.

Ontogeny. Developmental history of an organism.

Oocyte. The presumptive egg cell after the initiation of the growth phase of maturation. Syn., ovocyte.

Oogenesis. The process of maturation of the ovum; transformation of the oogonium to the mature ovum. Syn. ovogenesis.

Oogonia. The multiplication (mitotic) stage prior to maturation of the presumptive egg cell (ovum), found most frequently in the peripheral germinal epithelium.

Optic cup. Invagination of the outer wall of the primary optic vesicle to form a secondary optic vesicle made up of two layers; a thick internal or retinal layer continuous at the pupil and the choroid fissure, and a thin external layer which is pigmented. The cavity of the cup becomes the future posterior chamber of the eye.

Optic lobes. The thickened, evaginated, dorso-lateral walls of the mesencephalon.

Optic recess. A depression in the forebrain anterior to the optic chiasrna which leads to the optic stalks.

Optic stalk. The attachment of the optic vesicle to the forebrain, at first a tubular connection between the optic vesicle and the diencephalon. The lumen is later (8 day chick) obliterated by the development of optic nerve fibres.

Optic vesicle. Evaginations of forebrain ectodenn to form the primary optic vesicles which in turn invaginate to form the secondary optic vesicles or optic cups of the eyes.

Optkoel. The cavity of the primary optic cup.

Ora serrata. The line of separation between the retinal and lenticular zones of the eye cup.

Oral plate. Fused stomodeal ectoderm and pharyngeal endoderm to form the oral membrane. Breaks through to form the mouth. Syn., pharyngeal membrane, oral membrane, stomodeal plate.

Organizer. The chorda-mesodermal field of the amphibian embryo; a tissue area which has the power of organizing indifferent tissue into a neural axis; possibly comparable to Hensen's node of the chick embryo.

Ostium abdominale tubae. The most anterior, fimbriated end of the oviduct in female vertebrates; the point of, entrance of the ovulated egg into the oviduct; double in amphibia and on the left side only in the female bird. Syn., infundibulum of the oviduct.

Otocyst. The original auditory vesicle appearing at the level of the rhombencephalon in the amphibian embryo just before hatching and in the chick embryo at about the 16 somite stage, forming first as a placode. Syn., auditory vesicle.

Otolith. Granular concretion found within the (embryonic) ear.

Oviduccal membranes of ovum. Tertiary membranes applied over the egg as it passes through the oviduct, in both amphibia and chick.

Oviducts. The paired Muilerian ducts in both males and females, which generally degenerate in the males (except in amphibia).

Ovigerous cords. Columns or strands of tissue which divide the germinal epithelium of the prithordium of the ovary carrying primordial germ cells with them and later breaking up into nests of cells each of which contains an oogonium. Syn., egg tubes or cords of Pfluger (mammal).

Oviposition. The process of laying eggs.

Ovulation. The release of the egg from the ovary, not necessarily from the body.

Ovum. Latin for egg.

Paedogenesis. Reproduction during larval stage; precocious sex development.

Parthenogenesis, artificial. Initiation of development of an egg by artificial means.

Parthenogenesis, natural. Maturation of eggs of some forms leads directly to development without the aid of spermatozoa.

Parthenogenesis. Development of an egg without benefit of spermatozoon.

Parthenogenetic cleavage. Fragmentation of protoplasm of old and unfertilized chick eggs, originally thought to be true cleavage.

Parturition. The process or act of giving birth.

Path, penetration. The initial direction of sperm entrance into the egg, often shifting toward the egg nucleus along a new copulation path. Syn., entrance path.

Pecten. An aggregation of blood vessels embedded in mesenchyme, exclusively found in the bird eye about the fourth day of development, forming a ridge of tissue projecting into the optic cup through the proximal end of the chooid fissure. Becomes folded and fan-shaped by the 8th day.

Penis. Elongated genital tubercle, enclosed in genital folds (prepuce) and associated with genital swellings (scrotal pouches) in male mammal; organ of transfer of genital products from the testes of the male to the vaginal cavity of the female. Syn., phallus.

Peristomial mesoderm. Mesoderm of the amphibian gastrula derived from the (ventral) lips of the blastopore; in Amphioxus derived from the substance of the germ ring; in the chick derived from the primitive streak-itself. Opposed to gastral mesoderm.

Peritoneal cavity. The body cavity (coelom) separated from the pleural cavity by the pleuro-peritoneal septum including the septum transversum.

Peritoneum. Coelomic mesothelium of the abdominal region reinforced by connective tissue.

Perivitelline space. The space between the vitelline (fertilization) membrane and the contained egg, generally filled with a fluid. In the mammal it is the space between the zona radiata and the egg.

Peter's ovum. One of the rare specimens of a very early human embryo, from which a great deal of information has been gathered. '

Pfluege. Cords of the ovigerous layer which grows into the stroma of the mammalian ovary as ovigerous cords, carrying primitive ova with them.

Placenta. An extra embryonic vascular structure of placental mammals which serves as an organ of nutritive and respiratory exchange between the fetus and the mother.

Plasmosome. A true nucleolus.

Poikilothermal. Adjective meaning cold-blooded; animals whose body temperatures are subject to environmental changes because they lack regulating mechanisms. Opposed to homoiothermal.

Polar body. Relatively minute, discarded nucleus of the maturing oocyte (generally three). Syn., polocytes.

Polar. Pertaining, in most cases, to the animal pole although may refer to the vegetal pole, or both.

Pole, animal. Region of the egg, where the polar bodies are eliminated; ectoderm forming portion of the pre-cleaved egg. Syn., apical or animal hemisphere.

Pole, vegetal. Region of the egg opposite the animal pole; region of lowest metabolic rate; pole with greater density of yolk in telolecithal eggs; generally the endoderm forming region of the egg.

Polyembryony. Production of several separate individuals from one egg by an early separation of its blastomeres possible origin of some identical twins.

Polyestrous cycle. Reproductive cycles which occur at least several times a year (e.g., pig).

Polyestrus. Mammals which have more than a single estrus cycle in one year.

Polyinvagination. Theory of Pasteels that individual (endoderm) cells of the chick blastoderm migrate beneath the surface of the blastodisc and spread out to form the second of the primary germ layers.

Polyspermy. Insemination of an egg with more than single sperm generally occurring in the chick egg although hut a single sperm nucleus is functional, in syngamy.

Postestrus. Short period of regressive changes in the uterine mucosa following fruitless preparation for pregnancy.

Pre-migratory germ cell. Yolk laden cells of splanchnopleuric origin which migrate by way of blood vessels to the gonad primordia. Believed by some to be the precursors of gonad stroma and/or functional germ cells.

Preformation. Theory that the adult is represented in miniature within the egg or sperm, and that development is simply enlargement.

Pregnancy. Condition of actually bearing an embryo or fetus within the uterus or Fallopian tubes.'

Primary oocyte. The termination of the growth phase in the maturation of the ovum from the oogonial stage, prior to any maturational divisions.

Primary spermatocyte. Stage in spermatogenesis whose division results in secondary spermatocytes; stage beginning with growth of the spermatogonia.

Primitive folds. The uplifted sides of the primitive streak of the chick possibly corresponding to the lips of the amphibian blastopore.

Primitive groove. A groove through the center of the primitive streak, bounded by the primitive folds and terminated anteriorly by the primitive pit and posteriorly by the primitive plate.

Primitive gut. The space between the endoderm and the yolk of the chick embryo, derived from part of the subgerminal cavity. Syn., archenteron or primitive intestine.

Primitive ova. Oogonia within the ovary which multiply by mitosis without growth, in the mammalian ovary found along the cords of Pfluger.

Primitive pit. The anterior end of the primitive groove, sua rounded by Hensen's node, and probably representing the homologue in the chick of the amphibian neurenteric canal.

Primitive plate. The expanded posterior end of the primitive streak of the chick embryo, possibly corresponding to the sickle groove.

Primitive streak. The first sign of embryonic differentiation in the chick blastoderm; the material out of which the notochord and mesoderm originate anteriorly; possibly the homologue of the fused lateral lips of the amphibian blastopore. The embryo develops anteriorly to but not in direct line with the primitive streak. The streak first appears at about 18 hours in the chick.

Primordial germ cell. The diploid cells which are destined to become germ cells. Syn., primitive germ cells. E.g., oogonia and spermatogonia.

Proamnion. The area pellucida immediately anterior and lateral to the head process of the early chick embryo which is devoid of mesoderm for some time; consists only of ectoderm and endoderm.

Proctodeurn. An ectodermal pit in the region of the future cloaca which invaginates to fuse with hindgut endoderm to form the anal or proctodeal plate, later to rupture and form the anus.

Pronucleus. The egg nucleus after polar body formation and the sperm nucleus after entrance of the spermatozoon into the egg.

Pupil. The opening into the secondary optic vesicle, occluded in part by the lens, and regulated in diameter by the ciliary muscles of the iris.

Rauber's cells. The remnant of the mammalian trophoblast at the point of junction with the embryonic knob.

Recapitulation theory. Theory that embryonic development reviews the major steps in evolutionary history.

Regeneration. Repair or replacement of lost part or parts, a power gradually lost in the ontogeny of most animals. -

Regulation. A reorganization toward the whole; the power of pie-gastrula embryos to utilize materials remaining, after partial excision, to bring about normal conditions; more flexible power than regeneration,

Rete cords. Strands of epithelial cells, containing many primordial germ cells which connect with the seminiferous tubules and later become the vasa efferentia. In the bird. Syn., rete testis.

Retinal zone. Ectodermal derivatives of the optic cup consisting of the internal limiting membrane, retinal and lenticular zones, and outer pigmented layer. The retina proper includes portions from internal limiting membrane to the rods and cones inclusive.

Rutting period. Period of heightened sexual desire on the part of some male mammals, which period generally coincides with sexual receptivity on the part of the female.

Rutting season. Brief period of sexual activity occurring in some males; term used by animal breeders.

Saccus vaginalis. Scrotal pouch, peritoneal pocket in scrotal pouch.

Sach's law. All cells tend to divide into equal parts and each new plane of division tends to intersect the preceding one at right angles.

Sagittal. A mesial plane, or any plane parallel to it, dividing the right parts of the body from the left. Right angles to both the frontal and transverse planes.

Scierotome. Loose mesenchymal cells proliferated off from the inner and ventral edges of the myotomes (5 min frog) which contributes to the formation of the axial skeleton. In the chick (96 hours) three parts are distinguished the narrow undivided perichordal part; the dense aggregations of caudolateral cells; and the cephalic portion, all of which contribute to the axial skeleton.

Secondary oocyte. The stage in oogenesis between primary oocyte and ovum, may be either haploid or diploid, depending upon species considered and which maturation division occurs first.

Secondary spermatocyte. The stage in spermatogenesis whose next division results in haploid spermatids, these spermatocytes being either haploid or diploid depending upon species considered.

Segmentation cavity. The cavity of the blastula; in the chick it appears as result of cleavage parallel to the surface of the blastoderm, but the cavity never extends beneath the marginal cells of the periblast. Syn., subgerminal cavity, blastocoel.

Segmentation. Repetition of structural pattern; used as synonym for cleavage as well as for metamerism.

Semen. Mixture of secretions from the bulbo-urethral glands (Cowper's glands), prostate gland, seminal vesicles and the suspended spermatozoa.

Semination. The act of fertilizing, by the discharge of spermatozoa.

Semipbcenta. Type of placenta in which the uterine mucosa and the chorion do not actually grow together so that there is no tearing at birth. Syn., contact placenta.

Sero-amniotic connection. The final point of fusion of the various amniotic folds as they converge above the chick embryo, the knot of tissue remaining throughout the entire embryonic period of the chicks which modifies the formatting and disposition of the embryonic membranes. Original ectodermal knot becomes displaced by mesoderm which later develop a lumen allowing the connection between the albumen sac and the amniotic cavity (11th day). Syn., sero-amniotic raphe.

Serosa. syn., chorion.

Sertoli cell. Derivative of the sexual cords of the testis, found within the seminiferous tubule and functionally similar to the follicle cell in the ovary in that it is the nutritive, supporting, or nurse cell of the maturing spermatozoa. The heads of adult spermatozoa may be seen embedded in the cytoplasm of Sertoli cells.

Sex cell cord. Division of the sex-cell ridge or gonad primordium, not to be confused with sexual (rete) cords.

Sex determination. Generally means either the conditions of fertilization which determine the ultimate sex of the embryo, or the predetermination of sex by experimental means.

Sexual cycle. Periodic sequence of changes in the uterine mucosa of the female placenta. Mammals, apart from pregnancy, include uterine changes as follows: Diestrus-period of quiescence, if short.; Anerus-period of quiescence, if long.; Proestrus-period of construction, destruction, and some repair.; Estrus-period of repair. The cycle related to ovulation and to endocrine cycle.

Shell gland. The glandular portion of the bird uterus which provides the materials of the shell.

Shell membrane. Double membrane, consisting of a thick outer and thin inner layer of matted organic fibres crossing each other in all directions, the two membranes separated only by the air space at the blunt end of the egg.

Shell. A calcareous covering deposited on the chick (bird) egg during the 12-16 hour period it remains in the uterus, made up of carbonates and phosphates of calcium and magnesium, structurally consisting of an inner mammillary layer of calcareous particles. Intemediate spongy layer of mated calcareous strands. Outer, surface cuticle, which is porous, but otherwise structureless.

Sinus rhomboidalis. The region of the receding primitive streak around which the posterior ends of the neural folds are diverted in the chick embryo.

Sinus terminalis. An encircling blood vessel which appears at the outer margin of the area vasculosa of the chick embryo before the end of the first day, found also in eggs o marsupials although no yolk is present. Syn., versa terminalis.

Sinus venosus. The point of fusion of vitelline veins of the amphibian embryo or the most posterior of the original four chambers of the (chick) heart at 33 hours, bilaterally symmetrical and related to the ducts of Cuvieri and the ductus venosus to 45 hours, right horn elongated and left reduced with shifting of sino-auricular aperture (60-70 hours) and at 96 hours has the shape of horse shape between atrium and the septum transversum. The most anterior part involved in formation of right auricle.

Somatic umbilicus. A short thick, hollow stalk which connects the chick embryo with the underlying yolk-sac and the extra-embryonic membranes, composed of ectoderm and somatic mesoderm continuous with the amnion, umbilicus closed just before the time of hatching after the remnants of the yolk have passed into the gut.

Somatopleure. The layer of somatic mesoderm and closely associated ectoderm, the extension of which (from the body wall) gives rise to both the amnion and chorion.

Somite. Blocks of paraxial mesoblast metamerically separated by transverse clefts, derived from enterocoelic or gastral mesoderm and giving rise to the dermatome, myotome, and sclerotome.

Spawning. The act of expelling eggs from the uteri of anamniota (e.g., amphibia).

Sperm. The germ cell characteristically produced by the male. Syn., spermatozoon, sperm cell, male gamete, spermatosome.

Spermatic. The products of the second maturation division in spermatogenesis, the spermatids having certain cytological characteristics and being invariably haploid, cells which go through a metamorphosis into functionally mature spermatozoa.

Spermatocyte. Stages in spermatogenesis between the times the primordial germ cell (spermatogonium) begins to grow, without division, until after the division which results in spermatids.

Spermatogenesis. The entire process which results in the maturation of the spermatozoon.

Spermatogonium. The primordial germ cell of the male gonad, indistinguishable from somatic cells, both of which are diploid, stage prior to maturation when the presumptive spermatozoon undergoes rapid multiplication by mitosis.

Spermatophore. Sperm-bearing bundle, such as are shed by male urodeles, the bundles later to be picked up by cloacal lips of the female.

Spermatozoon. The functionally mature male gamete. Syn., sperm.

Spina bifida. Split tail, generally involving the spine, in developing embryo caused by a variety of environmental conditions, most of which act through interference with normal gastrulation and neurulation.

Spiracle. Short funnel between the body wall and the operculum on the left side of the head of the frog tadpole, the only exit for water passing through the gill chambers to the exterior.

Splanchnic mesoderm. The visceral mesoderm, or that nearest the embryonic axis in the lateral plate.

Splanchnic. Refers to the viscera, opposed to somatic or body.

Splanchnocranium. That portion of the skull which is preformed in cartilage and which arises from the first three pairs of visceral arches. Opposed to neurocranium.

Splanchopleure. The layer of endoderm and inner mesoderm (splanchnic) within which develop the numerous blood vessels of the area vasculosa and later the yolk-sac septa, the layers within the body which give rise to the lining and to the musculature of the alimentary canal.

Stereoblastula. Solid blastula as found in Crepidula.

Stratum granulosum. Layer of follicle cells surrounding the mammalian ovum.

Sub-cardinal veins. Embryonic veins ventral to the nephric tissue the posterior portions of which fuse to contribute to the formation of the inferior vena cava in the chick.

Sub-claviin veins. These arise primitively as branches of the posterior cardinals, receiving blood (in the chick) from the wings and walls of the thorax.

Sub-notochordal rod or bar. The hypochordal rod of the amphibian embryo or the point of fusion of the caudal sclerotomal (cephalic vertebral) components beneath the notochord of the chick embryo.

Sub-zonal layer. The morula of the mammalian blastocyst lies within a sub-zonal layer of cells later called the trophoblast.

Substantia propria. Mesenchyme of the cornea continuous with the sclera.

Supernumerary nuclei. Nuclei of excess spermatozoa which invade the chick egg at fertilization, many of which cause accessory cleavages and then degenerate.

Sustentacular cell. A cell which provides nourishment for another, such as the Sertoli or follicle cells of the gonads.

Syncytial trophoderm. Layer of trophodermal cells of the human embryo outside of but probably derived from the cell layer of Langhans.

Synezesis. The stage in maturation between synaptene and pachytene when the chromatin threads are short and thick and the ends away from the centrosome are tangled.

Syngamy. Specifically the fusion of the gamete pronuclei, but also the union of gametes at fertilization. Syn., zygote-genesis, fertilization.

Tail bud. Concurrent with the development of the tail fold a mesodermal tail bud mass may be distinguished, at 48 hours in the chick embryo.

Tail fold. A sulcus begins to develop beneath the posterior end of the chick embryo at about 48 hours, giving rise to a tail fold similar to the head fold except that from the beginning it is made up of ectoderm and somatic mesoderm. In mammals the tail fold often appears prior to the head fold and is longer. Syn., amniotic tail fold.

Telobiosis. Fusion of embryos end-to-end.

Teratology. The stud) of the causes of monster formation.

Tertiary membranes. The soft membraneous and hard calcareous coverings of the chick egg, deposited on the egg within the uterus.

Theca externa. The outermost of the coverings of the ovarian follicle, rather loose connective tissue with abundant blood supply. Continuous with ovarian stroma.

Theca folliculi. Refers to membranous and cellular coverings of the ovum within the ovary.

Theca interna. The layer of connective tissue consisting of closely packed fibres, possibly some of smooth muscle, immediately external to the ovarian follicle of birds and mammals. Less vascular and more compact than theca externa.

Theca. Connective tissue covering, generally refers to coverings of ovarian follicle.

Tissue culture. In vitro culturing of isolated tissues, the excision of tissues or organs and their maintenance in an artificial medium, generally consisting in part of embryonic extracts or blood plasma.

Torsion. The twisting of the (chick) embryo so that it lies on its (left) side, caused in part by the passive resistance of the underlying inert yolk plus the development of the various cephalic flexures.

Totipotency. Related to the theory that the isolated blastomere is capable of producing a complete embryo.

Transplant. An embryonic area (cell, tissue, or organ) removed to a different environment.

Trophectoderm. Region of continuity of ectoderm and outer layer of mammalian trophoblast, extra-embryonic ectoderm following gem layer differentiation.

Trophoblastic villi. Finger-like projections of chorionic mesoderm comprising the trophoderm of mammalian embryos.

Trophoderm. Trophectoderm reinforced by a layer of somatic mesoderm. Syn., extra-embryonic somatopleure, serosa.

True placenta. A placenta in which the chorion and uterine mucosa are intimately associated, in contrast with the primitive contact type. Syn., burrowing placenta.

Umbilical arteries. Branches of the sciatic arteries of the chick embryo which appear on the 4th day, the right member being the smaller, carry blood to the allantois.

Umbilical veins. At first paired veins in the lateral body wall of the chick embryo (4th day) which bring blood from the allantois and join the ducts of Cuvier, right vein disappearing and the left changing its connection to join the anterior half of the ductus venosus. At hatching, only the proximal portion persists as a vein in the ventral body wall.

Umbilicus. The stalk like connection between the embryo and all extra-embryonic structures, including the somatic stalk, allantoic stalk plus its arteries and veins, and the yolk stalk with its arteries and veins.

Uterine milk. A viscid fluid secreted by the uterine mucosa consisting of fats, proteids and glycogan, a source of embryonic nutrition in mammals.

Vagina. Short, thin-walled cavity which opens from the uterus into the cloaca of the bird, cavity within which the hen's egg is retained after 4 p.m. until the following morning, but from the walls of which no contributions are received, the cavity of the female mammal possessing, at its external boundaries, a homologue of the male penis called the clitoris.

Vasa deferentia. Mesonephric or Wolffian ducts of the frog, which ducts persist as the male gonoducts of the bird and mammal, connecting with the testes through the vase efferentia and epididymis and functioning as sperm ducts after the degeneration of the embryonic megonephros and the development of the gonads. Sing., vas deferens.

Vasa efferentia. The ducts which convey (frog) sperm from the collecting tubules through the mesorchium to the malphigian corpuscles of the mesonephric kidney, derived (in the chick) from rete cords and connected with the mesonephric tubules of the anterior (sexual) half of the mesonephric or Wolffian body which become the epididymis. Sing., vas efferens.

Vegetal pole. The pole of a telolecithal egg where there is the greatest concentration of yolk, generally opposite the animal pole and the location of the germinal vesicle, Syn., vegetal or vegetative hemisphere, abapical, or antipolar hemisphere.

Velar plate. Folds or flaps developing anterior and posterior to the branchial regions of the frog (anuran) embryo derived from the pharyngeal wall and serving as a gross sifting organ between the pharynx and the gill (branchial) chamber.

Vena porta sinistra. Some of the veins from the gizzard and proventriculus of the chick embryo which enter the left lobe of the liver.

Vertebra. Derivatives of the sclerotome which surround the nerve cord and notochord, and finally incorporate the notochord by chondrification and ossification (centrum).

Villus. A finger-like projection, such as the chorionic villus which is a projection of the mammalian chorion into folds of the uterine mucosa.

Vitelline artery. Paired oraphalomesenteric vessels which, in the chick, later fuse (as the dorsal mesentery forms) to go out through the umbilical stalk to the yolk as the vitelline (yolk) arteries, originating from the dorsal aorta.

Vitelline membrane. Delicate, outer, non-living egg covering derived while the egg is still within the ovary, probably by joint action of the egg and its follicle cells, successful insemination.

Vitelline substance. Yolk.

Vitelline. Pertains to yolk. (E.g. vitelline vein brings blood from the yolk, vitelline membrane is that which covers the yolked egg).

Vitreous humor. The rather viscous fluid of the eye chamber posterior to the lens, formed by cells budded from the retinal wall and from the inner side of the lens, hence ectodermal and probably also mesenchymal in origin.

Viviparous. Animals which bring forth young in advanced state of development, more advanced than eggs. (E.g., humans)

Wing bud. Mesenchyme, probably of sclerotomal origin, which forms paired buds at the level of the somites #I4 to #16 by the 4th day of chick development the primordia of the wings and possibly parts of the pectoral girdle.

Yolk nuclei. Darkly staining chromatin-like substances within the cytoplasm of the young (immature) eggs around which the yolk is accumulated during the growth phase of oogenesis. May be derived from nucleoli which escape from the nucleus.

Yolk plug. A plug formed by large yolk cells which are too large to be immediately incorporated in the floor of the archenteron of the amphibia embryo, hence are found protruding slightly from the blastopore. Size of the plug is often used to determine the approximate stage of gastrulation.

Yolk sac endoderm. The endoderm of the splanchnopleure surrounding the yolk itself, which develops glandular structures within the septa for digestion of the yolk at points of contact.

Yolk sac placenta. Type of Marsupial placenta in which the blood vessels are derived from the area vasculosa of the yolk sac. (E.g. DASYURUS).

Yolk sac septa. Deep folds of splanchnopleure which surrounds the yolk the folds projecting into the yolk to accomplish digestion by means of glands which secrete enzymes.

Yolk sac umbilicus. The region where the yolk of the chick egg is not surrounded by the yolk sac until the development of the albumen sac is complete.

Yolk sac. Extra-embryonic splanchnopleure which grows around and finally all but encircles the yolk mass, remaining connected with the mid-gut by the yolk-stalk finally taken into the mid-gut through the stalk just before hatching. Function is to provide extensive surface area for absorption of nutriment.

Yolk stalk. Thick-walled stalk which connects the midgut and the yolk-sac of the developing chick embryo, obvious at 5 days.

Yolk. Highly nutritious food (metaplasm) consisting of non-nucleated spheres and globules of fatty material found in all except alecithal eggs. Syn. lecithin.

Zona pellucida. Transparent, non-cellular, secreted layer immediately surrounding the mammalian ovum, corresponding to the vitelline membrane of lower forms.

Zona radiata. Zona pellucida which exhibits radial striations, not to be confused with the corona radiata. Syn., zona pellucida of the mammalian egg.

Zonary placenta. The placenta of Carnivores in which the villi are arranged in bands or zones.

Zone of overgrowth. Narrow superficial rim of marginal periblast cells of the early chick embryo beyond the zone of junction which represents epibolic growth out over the unsegmented yolk. Syn., margin of overgrowth.

Zones of junction. Portion of marginal periblast of chick embryo which is not separated from the germ wall beneath by any cytoplasmic cleavages, i.e. region of syncytium, peripheral to which is the zone of overgrowth. Syn., region of the germ wall.

Zygote. The diploid cell formed by the union of two gametes. Syn., fertilized egg.

Collateral Readings

Agarwal, S.K., and Niazi, I.A., 1977. Normal table of developmental stages of the Indian bullfrog, Rana tigrina Daud. (Ranidae, Anura, Amphibia). Proceedings of the National Academy of Sciences, India 47 (B) 79, 92.

Allen, B.M. 1929. The influence of the thyroid gland and hypophysis upon the growth and development of amphibian larvae Exuart. Rev. Biol. Vol. 4, pp. 335.

Arey, I.B. 1954. Developmental Anatomy. W.B. Saunders Co., Philadelphia.

Assheton, R. 1899. The development of the pig during the first ten days. Quart. J. Micr. Sci. 4: 329-359.

Balinsky, B.I. 1960. An introduction to embryology. W.B. Saunders Co. Philadelphia.

Balinsky, B.J. 1981. An Introduction to Embryology. Saunders College Publishing, Philadelphia.

Ballard, K. J. and Holt, S.J. 1968. Cytological and cytochemical studies on cell death and digestion in foetal rat the role of macrophages and hydrolytic enzymes. J. Cell Sci., 3: 245-262.

Barth, L.G. 1953. Embryology, Dryden Press, New York.

Bellairs, R. 1971. Developmental. Processes in Higher Vertebrates. Logos Press, London. xiv + 366 pp.

Berrill, N.J. 1971. Developmental Biology. McGraw-Hill Book Co., New York. vii + 535 pp.

Berrill, N.J. and Karp, G. 1976. Development. Tata McGraw-Hill Pub. Co., New Delhi.

Bodemer, C.W. 1968. Modern Embryology. Holt, Rinehart and Winston, Inc., New York.

Briggs, R. and King, T.J. 1952. Transplantation of living nuclei from blastula cells into enucleated frogs eggs. Proc. Nat. Acad. Sci. 38: 455-463.

Cairns, J.M. and Gasseling, M.T. 1957. The role of the apical ridge of ectoderm in the differentiation of the morphological structure and inductive specificity of limb parts in the chick. J. Morphol. 101: 57-87.

Croisille, Y. and LeDouarin, N.M. 1965. Development and regeneration of the liver, in DeHaan and Ursprung, Organogenesis. Holt, Rinehart and Winston, Inc., New York. pp. 421-466.

DeHaan. Robert, L. and Heinrich Ursprung, eds. 1965. Organogenesis. Holt Rinehart and Winston, New York.

Dent, J.N. and Hunt, E.L. 1954. Radiotracer techniques in embryological research. J. Cell. Comp. Physiol. suppl. 43. 1: 77-102.

Derrick. G.E. 1937. An analysis of the early development of the chick by means of the mitotic index. J. Morphol. 61: 257-284.

Dutta S.K. and Mohanty-Hejmadi, P., 1976. Breeding and life history of the Indian bull frog Rana tigerina. Prakruti-utkal. Willy. J. of Sci. 13 (1 and 2): 51-59.

Etkin, W. 1955. Metamorphosis: In Willier, Weissan Hamburger (eds.). Analysis of Development: W.B. Saunders Co. Philadelphia.

Etkin, -.W. 1968. Hormonal control of amphibian metamorphosis. In: metamorphosis, a problem in Developmental Biology (Eds. W.B. Etkin and L.I. Gilbert). pp. 313-348. Meredith ed. Lowa

Frieder, E. 1963. The chemistry of amphibian inetamorphosis. Sc. Ann. 209: 110-118.

Gabiel, ivt,L. and Fogel, S. 1955. Great experiments in Biology. Prentice-Hall, Inc.,

Englewood cliffs, New Jersey.

Gilbert Scott, F. 1991. Developmental Biology (3rd ed.), Sinauer Associates, Inc. Sunderland Massachusetts. 25, Gilchrist*, F.G. 1968. A survey of Embryology. McGraw Hill Book Co., New York.

Glucksmann, A. 1951. Cell death in normal vertebrate ontogeny. Bil. Revs. Cambridge Phil. Sod. 26: 59-86.

Goel, S.C. 1983. Laboratory Exercises in Developmental Biology. Indian Society of Developmental Biologists. Dept. of Zoology, Univ. of Poona (India).

Goel, S.C: and Muthur, J.K. 1978. Cell death in reptilian limb morphogenesis. Ind. J. exp. Biol., 16: 653-655.

Gosner, K.L. 1960. A simplified table. for staging anuran embryos and larval with notes on identification. Herpatologica, 16: 183-190.

Gurdon, J.B. 1962. The developmental capacity of nuclei taken from intestinal epithelium cells of feeding tadpoles. J. Embryol. Exp. Morph. '10:622-64.0.

Hamburger, V. 1966. A Manual of Experimental Embryology. University of Chicago Press, Chicago. pp. xvii + 221.

Hamburger, V. and Hamilton, H.L. 1951. A series of normal stages in the development of the chick embryo. J. Morph., 88: 49-92.

Hamilton, W.J., Boyd, J.D. and Mossmami, H.W. 1962. Human Embryology. E. Heffer and Sons Ltd. Cambridge, England.

Harrison, J.R. 1954. Growth and differentiation of the embryonic chick eye in vitro. I. On yolk albumen fractions obtained with centrifugation and heat J, Exp. Zool., 127: 493-510.

Heller, C.G. and Clermont, Y. 1963. Spermatogenesis in man. An estimate of its duration. Science, 140: 184-186.

Hinchliffe, J.R. and Ede, D.A. 1973a Cell death and the development of limb form and skeletal pattern in normal and wingless (ws) chick embryos. J. Embryol. exp. Morph. 30: 753-772.

Huettner, A.F. 1949. Fundamentals of comparative Embryology of the vertebrate's Rev. ed. Macmillan, New York.

Humason, G.L. 1972. Animal Tissue Techniques. W.H. Freeinan and co., San Francisco.

Huxley, J.S. and deBeer, G.R. 1934. The Elements of Experimental Embryology. Cambridge Univ. Press, London.

Jangir, O.P. and Niazi, I.A. 1978. Stage dependent effects of vitamin A excess on limbs during ontogenesis and regeneration in tadpoles of Bufo nzelanostietus. Ind. J. Exp. Biol. 16: 438-445.

Jangir, O.P. and Niazi, I.A. 1979. Increased mitotic activity and precocious re-differentiation in regenerating hind limbs of Bufo melanostictus tadpoles exposed to thyroxine after amputation: Proc. 66th session Ind. Sci. Cong. Ass. Hyderabad.

Jangir, O.P. and Niazi, I.A. 1979. Regeneration of whole limb from shank level blastema grafted auto plastically in the orbit in the toad tadpoles immersed in Vitamin A solution following first amputation. Proc. All India Symposium on Exptl. Zool. Jaipur.

Jangir, O.P., Digvijay Singh and Goswami, M.K. 1995. Study on regenerative ability of lens in Post-natal mice under the influence of vitamin A. XI Nat. Symp. on Develop. Biol. Rohtak.

Jangir, O.P., Digvijay Singh and Sushma Garg, 1995. The fate of regenerating blastema implanted into enucleated eyes of tadpoles of Rana cyanopitlyetis under the influence of vitamin A. XI National Symp. on Developmental Biol. Rohtak.

Jangir, O.P., Digvijay Singh and Swami, K.K. 1995. A study of lens regeneration in the tadpoles of Rana cyanopitlycns under the influence of vitamin A. XI Nat. Symp. on Develop. Biol. Rohtak.

Kaltenbach, J.C. 1953. Local action of thyroxin on amphibian metamorphosis. II Development of eye lids, nictitating membrane cornea, and extrinsic ocular muscles in Rana pipiens larvae effected by throxincholesterol implants. J. Exptl. Zool. 122, 41-51.

Kataeiri, C. 1972. A method for assaying the, jell}, digesting activity of the hatching enzyme from frog embryos. J. Fac. sci. Hokkaido Univ. ser. vi, Zool. 18, 461-468.

Kemp. N.E. 1956. Electron microscopy of growing oocytes of Rana pipiens. J. Biophys. Biochern. Cytol. 2: 281-292.

Khan, M.S. 1965. A normal table of *Bufo melanostictus* (Schneider). Biological: 1-39 (1965).

Kollros, J.J. 1961. Mechanism of amphibian metamorphosis: Hormones Am. Zol. 1: 107-114.

Lewis, W.H. 1904. Experimental studies on the development of the eye in amphibia. I. On the origin of the lens. Am. J. Anat. 3: 505-536.

Lillie, F.R. 1952. The Development of the Chick. Henry Holt, New York.

Loomis, W.F. 1958. Foundations of embryology. McGraw-Hill Book Co., New York.

Mathews Willis, W. 1972. Atlas of Descriptive Embryology. The Macmillan Co. 866 Third Avenue, N.Y., New York 10022.

Mazia, D. 1961. How cell divide. Sci. Am. 205, 100-120.

McEwen, R.S. 1949. Vertebrate Embryology. Holt, Rinehart and Winston, Inc., New York. 3D ed., xv + 699 pp.

McEwen, R.S. 1957. Vertebrate Embryology. Holt, Rinehart and Winston Inc. New York.

McEwen, R.S. 1969. Vertebrate Embryology, 4th Ed. Oxford and IBH Publishing Co. New Delhi.

McKenzie, J. 1976. An Introduction to Developmental Biology. Blackwell Scientific Publication, Oxford, England.

Menkin, M.F. and Rock, J. 1948. In vitro fertilization and cleavage of human ovarian eggs. Am. J. Obst. and Gynecol., 55: 440-452.

Mohanty-Hejmadi, P., Dutta and Sushil, K. 1980. Breeding and development of Rana cyanophlyctis Schneider. J. Bombay Nat. Hist. Soc. 76: 291-296.

Mohanty-Hejmadi, P., Nayak, B.K. and Kanungo, Jyotshnabala, 1983. The reproductive biology of the Indian skipper frog Rana cyanophlyctis. Herpetological Reviews SSA, USA (In press).

Mohanty-Hejmado, P. 1977a. Care and maintenance of Indian frogs: Juveniles and adults. Prakruti-Utkal Univ. J. Science. 11: 75-79.

Mohanty-Hejmado, P. 1977b. Care and management of amphibian embryos. Prakruti-Utkal Univ. J. Science. 11: 81-87.

Mulherkar, L., Goel, S.C. and Joshi, M.V. 1977. Effects of cytochalasin H on the cleaving eggs of Limnaea acuminata. Ind. J. exp. Biol. 15: 1089-1093.

Naizi, S. 1978. Ontogeny of photomechanical movements of retinal pigment in response of light and dark in the toad Bufo melanostictus (Schneider). Indian J. exp. Biol. 16: 425-429.

Needham, J. 1931. Chemical Embryology. The Macmillan Co, New York. xxii + 2021 pp.

Nelson Olin, E. 1953. Comparative Embryology of the vertebrates Blakiston, New York.

New, D.A.T. 1966. The Culture of Vertebrate Embryos. Logos Press, Academic Press, London. pp. xi + 245.

Niazi, I.A. and Jangir, O.P. 1989. In recent trends in regeneration research Eds. V. Kiortsis, S. Koulssolakos, H. Wallace. Pienum New York. pp. 355-370.

Niazl, I.A., Jangir, O.P. and Shivpal, 1977. Thyroid hormone a causative factor in the loss of limb regeneration capacity in anurans. Proceedings of the first symposium of the Indian Society of Developmental Biologists.

Noble, G.K. and Brady, M.K. 1930. The courtship of the plethodontid salamanders, Copeia,

N.Y., 52-54.

Patten, B.M. 1957. Early Embryology of the Chick. McGraw-Hill Book Co., New York.

Patten, B.M., and Carlson, B.M., 1977. Foundations of Embryology, Tata McGraw Hill, New Delhi.

Pattern, B.M. 1968. Human Embryology. McGraw-Hill Book Co., New York.

Ramaswami, L.S. and Lakshman, A.B. 1959. The skipper frog as a suitable embryological animal and an account of the action of mammalian hormones on spawning the same. Proc. Inst. of Science, India 25: 68-79.

Raven, C.P. 1968. Morphogenesis: The Analysis of Molluscan Development. Pergamon Press, London.

Ris, H. 1955. Cell division. In. Willer, Weiss and Hamburger (eds.) Analysis of Development. W.B. Saunders Co., Philadelphia.

Rugh, R. 1953. The Frog. Mc-Grew-Hill, New York.

Rugh, R. 1962. Experimental embryology: Techniques and Procedures. 3rd ed. Burgess Publ. Co. Minneapolis.

Rugh, R. 1962: Experimental Embryology. Burgess publishing Co., Minneapolis.

Si. Rugh, R. 1968. Experimental Embryology. 3rd Ed. Burgess Publishing Company. Minneapolis.

Rugh, R. 1968. Experimental Embryology. Techniques and Procedure. Burgess Publ Co., Minneapolis. pp. ix + 501.

Saunders, J.W. 1948. The proximodistal sequence of origin of the parts of the chick wing and the role of the ectoderm. J. Exp. Zool. 108: 363-403.

Saunders, J.W. 1970. Patterns and Principles of Animal Development. The Macmillan Co. New York. 296 pp.

Saunders, J.W. Jr. 1982. Developmental Biology. The Macmillan Co., N w York.

Sedra, S.N. and Michael, M.I. 1961. Normal table of the Egyptian toad, Bufo regularis Reuss, with an addendum on the standardization of the stages considered in previous publications. Cesk. Morfol., 9 (4): 333-351.

Shivpal and Niazi, I.A. 1977. A table of the normal development of the toad, Bufo andersonii Boulenger, Rajasthan University studies in Zoology, Jaipur, India.

Shumway, W. 1940. Stages in the normal development of Rana pipiens. 1. External forms, Anat. Rec., 78: 139-147.

Shumway, W. and Adamastone, F.B. 1954. Introduction to Vertebrate Embryology. John Wiley and Sons, New York.

Spemann, H. 1928. Die Entwicklung seitlicher and dorso-ventraler Keimhalften bei verzogetier Kernversorgung. Z. Wiss. Zool. 132: 105-134.

Spemann, H. 1938. Embryonic Development and Induction. Yale. University press. New Haven, Conn.

Sussman, M. 1960. Animal growth and development. Prentice-Hall, Inc. Englewood cliffs, New Jersey.

Tahara, Y. 1959. Table of the normal developmental stages of the frog Rana japonica. II. Later development Jap. J. Exptl. Morphol. 13.

Taylor, S.C. and Kollros, J.J. 1946. Stages in the normal development of Rana pipiens larvae, Anat. Rec., 94: 7-23.

Torrey Theodore, W. 1962. Morphogenesis of the Vertebrates. John Wiley and Sons, New York and. London.

Torrey, T.W. 1971. Morphogenesis of the Vertebrates. J. Wiley, Net York: Railed At 529 DO.

Waddington, C.H. 1956. Principles of Embryology. The Macmillan Co. New York, 3rd Printing, 1960, x + 510 pp.

Waddington, C.H. 1962. New Patterns in Genetics and Development. Columbia University Press, New York,

Waddington, C.H. 1966. Principles of Development and Differentiation. The Macmillan Co., New York. 101. West, L.S. 1951. The Housefly. Comstock Publ. Co., Ithaca, New York. Hewitt, C.G. 1914. The Housefly. University Press, Cambridge.

Whittaker, J. Richard, 1968. Cellular Differentiation. Dickenson, Belmont, California.

Willer, B.H., Weiss, P.A. and Hamburger, V. 1955. Analysis of development. W.B. Saunders Co., Philadelphia.

Wilier, B.H. and Oppenheimer (eds.) 1964. Foundations of Experimental Embryology. Prentice-Hall, Inc., Englewood Cliffs.

Witschi, E. 1956. Development of Vertebrates. W.B. Saunders Co. Philadelphia. xvi + 588 pp.

Wright, A.H. and Wright, A.H. 1949. Hand book of frogs and toads of the United States and Canada. Comstock Publ. Co. Ithaca, New York.

Zwilling, E. 1961. Limb morphogenesis. In Advances in Morphogenesis, (eds. M. Abercrombie and J. Brachet) pp. 301-330. New York, Academic Press.